THEORY AND PRACTICE OF COMPUTATION

T0136229

PROCEEDINGS OF THE WORKSHOP ON COMPUTATION: THEORY AND PRACTICE (WCTP 2019), 26 – 27 SEPTEMBER 2019, MANILA, THE PHILIPPINES

Theory and Practice of Computation

Edited by

Shin-ya Nishizaki
Tokyo Institute of Technology, Tokyo, Japan

Masayuki Numao
Osaka University, Osaka, Japan

Merlin Teodosia Suarez
De La Salle University, Manila, The Philippines

Jaime Caro
The University of the Philippines Diliman, The Philippines

CRC Press
Taylor & Francis Group
Boca Raton   London   New York

CRC Press is an imprint of the
Taylor & Francis Group, an **informa** business

A BALKEMA BOOK

CRC Press/Balkema is an imprint of the Taylor & Francis Group, an informa business

© 2021 Taylor & Francis Group, London, UK

Typeset by Integra Software Services Pvt. Ltd., Pondicherry, India

Library of Congress Cataloging-in-Publication Data

Applied for

Published by: CRC Press/Balkema
     Schipholweg 107C, 2316XC Leiden, The Netherlands
     e-mail: Pub.NL@taylorandfrancis.com
     www.routledge.com – www.taylorandfrancis.com

ISBN: 978-0-367-41473-3 (Hbk)
ISBN: 978-0-367-81465-6 (eBook)
DOI: 10.1201/9780367814656
https://doi.org/10.1201/9780367814656

Theory and Practice of Computation – Nishizaki et al (eds)
© 2021 Taylor & Francis Group, London, ISBN 978-0-367-41473-3

Table of contents

Preface

Modern research studies on computation started in the 1930s with the study of computability. At around the same time, computers were developed. In computer science, harmony of theory and practice have been considered important. Researchers in the field should establish a theory to address real world issues. On the other hand, experimentation and simulation are thought to be a trigger for improvement of theories.

WCTP 2019 is the nineth workshop organized by the Tokyo Institute of Technology, The Institute of Scientific and Industrial Research – Osaka University, University of the Philippines – Diliman and De La Salle University-Manila that is devoted to theoretical and practical approaches to computation.

It aims to present the latest developments by computer science researchers in academe and industry working to address computational problems that can directly impact the way we live in society.

Following the success of WCTP 2011–2018, WCTP 2019 was held in De La Salle University, on September 26 and 27, 2019.

This post-proceedings is the collection of the selected papers that were presented at WCTP 2019.

The program of WCTP 2019 consisted of selected research contributions. It included the most recent visions and researches of 17 contributions. We collected the original contributions after their presentation at the workshop and began a review procedure that resulted in the selection of the papers in this volume. They appear here in the final form.

July, 2020

Shin-ya Nishizaki
Masayuki Numao
Jaime Caro
Merlin Teodosia Suarez

Scientific committee

Jaime Caro (*University of the Philippines – Diliman*)
Rommel Feria (*University of the Philippines – Diliman*)
Henry Adorna (*University of the Philippines – Diliman*)
Francis Cabarle (*University of the Philippines – Diliman*)

John Paul Vergara (*Ateneo de Manila University*)
Mercedes Rodrigo (*Ateneo de Manila University*)

Allan A. Sioson (*Cobena*)
Jasmine Malinao (*Headstart Business Solutions*)
Annette Lagman (*UP System Information Technology Foundation*)

Merlin Suarez (*De La Salle University – Manila*)
Raymund Sison (*De La Salle University – Manila*)
Jocelynn Cu (*De La Salle University – Manila*)
Gregory Cu (*De La Salle University – Manila*)
Judith Azcarraga (*De La Salle University – Manila*)
Ethel Ong (*De La Salle University – Manila*)
Charibeth Cheng (*De La Salle University – Manila*)
Nelson Marcos (*De La Salle University – Manila*)
Rafael Cabredo (*De La Salle University – Manila*)
Joel Ilao (*De La Salle University – Manila*)
Marnel Peradilla (*De La Salle University – Manila*)
Ronald Pascual (*De La Salle University – Manila*)
Anish M.S. Shrestha (*De La Salle University – Manila*)

Rhia Trogo (*IBM Philippines*)

Koichi Moriyama (*Nagoya Institute of Technology*)

Masayuki Numao (*Osaka University*)
Ken-ichi Fukui (*Osaka University*)

Satoshi Kurihara (*Keio University*)
Mitsuharau Yamamoto (*Chiba University*)
Hiroyuki Tominaga (*Kagawa Univeristy*)

Shin-ya Nishizaki (*Tokyo Insitute of Technology*)
Takuo Watanabe (*Tokyo Institute of Technology*)
Masaya Shimakawa (*Takushoku University*)
Shigeki Hagihara (*Chitose Institute of Science and Technology*)

Theory and Practice of Computation – Nishizaki et al (eds)
© *2021 Taylor & Francis Group, London, ISBN 978-0-367-41473-3*

Organizing committee

Merlin Suarez (*De La Salle University – Manila*)
Judith Azcarraga (*De La Salle University – Manila*)
Rafael Cabredo (*De La Salle University – Manila*)

Acknowledgement

WCTP 2019 required a lot of work that was heavily dependent on members of the program committee, and lastly, we owe a great debt of gratitude to De La Salle University, for organizing the workshop.

Theory and Practice of Computation – Nishizaki et al (eds)
© 2021 Taylor & Francis Group, London, ISBN 978-0-367-41473-3

Updating a closed-world pharmacy information system to an open-world pharmacy information system in the Philippine context

Ruahden F. Dang-awan & L.L. Figueroa
Department of Computer Science, University of the Philippines Diliman, Philippines

ABSTRACT: The Pharmacy Information System (PIS) is a system designed to assist pharmacists in safely managing the medication process and support various activities in the pharmacy. (Alanazi et al., 2018) Systems created in the past, like pharmacy information systems (PIS) in the Philippines, were built with a closed-world assumption; they have yet to utilize open-world software technology and new software engineering techniques. Hence, for the PIS to be improved, it must be updated to adapt open-world settings.

The specific PIS that will be studied in this paper is *Medication Management System (MMS)*. MMS is a local PIS developed by a team of undergraduate students from the University of the Philippines Manila.

This paper shows that updating MMS, a closed-world PIS, to the open-world in the Philippine context is feasible by applying Philippines' Department of Health PHIE Lite as an interoperability standard; using MVC framework for modularization, statecharts for documentation of external behavior, and Parnas' module specification technique (Parnas, 1972b) for source code documentation. Additionally, this paper can also serve as a guide for other pharmacies and developers who want to update their PISs.

1 INTRODUCTION

A Pharmacy Information System (PIS) is usually a sub-system of a hospital information system (HIS). (Kazemi et al., 2016) The PIS is designed to assist pharmacists in safely managing the medication process; (Alanazi et al., 2018) it supports the distribution and management of drugs, it shows drug and medical device inventory, and facilitates the preparation of needed reports. (Kazemi et al., 2016) PISs also support other activities in the pharmacy including inpatient and outpatient order entry, management and dispensing, inventory and purchasing management, clinical monitoring, intervention management, pricing, charging, and billing, and administration of medication. (Alanazi et al., 2018).

The potential benefits of PIS on patient care are directly related to its impact on communication, outcomes, and medication management process. (Alanazi et al., 2018) Its implementation can reduce overall medication errors related to picking, preparation, and administration of drugs in the intensive care unit. (Chapuis et al., 2010) Like any technology, the PIS is required to evolve to fulfill the needs of patients and assist pharmacists.

A study by Lucianio Baresi *et al.* (Baresi et al., 2006) discussed how systems in the past have been built with "closed-world" assumptions. These closed-world assumptions assume that the boundary between system and environment is known and unchanging. In the context of PISs, this means that the system assumes that inputs will forever be of the same type and values, that there is no need for connectivity since there are no other services needed in the system except the built-in ones. However, this assumption is no longer applicable on today's open-world setting. The current generation of applications need techniques that let software react to changes by self-organizing its structure and self-adapting its behavior.

Hence, for the PISs to be improved, it could be updated to adapt to open-world settings. For the PIS, it means that we should assume there are inputs and services that will be needed outside the system. As a proof of concept, a PIS called *Medication Management System (MMS)*[1] developed for a local pharmacy will be used as an example.

This paper aims to be proof that updating a closed-world PIS to an open-world PIS in the Philippine context is possible; it also serves a guide for other pharmacies and developers who want to update their PISs.[2]

2 LITERATURE REVIEW

As of now, there are no guidelines on how to specifically update a closed-world PIS to an open-world setting. Hence, for this section, published literature related to PISs will be reviewed.

2.1 *Computers in pharmaceutical management*

The Managing Drug Supply (MDS) book (Embrey and for Health (Firm), 2013) is the leading reference on how to manage essential medicines in developing countries.[3] Chapter 50 of the book is entitled *Computers in pharmaceutical management.*

In chapter 50, the authors discuss the uses and special issues in using computers in pharmaceutical management, specifications for computer applications in pharmaceutical management, and requirements for maintaining and supporting computers. But for this subsection, we will be looking at the steps needed in creating custom software for pharmaceutical management:

1. Define the system requirements.
2. Choose the software and tools of developing the custom program.
3. Design a system.
4. Develop and program the system.
5. Test and debug the system.
6. Implement the system through data entry and training.
7. Develop system documentation and a complete user's manual.
8. Provide system support, revision, and upgrades.

This step-by-step list can serve as a guide for those who want to build a PIS or modify an existing PIS. The modification of an existing PIS is a part of step 8, but after every modification, the developers must repeat steps 5-8.

MMS, the current PIS being studied, lacks documentation and a user's manual. This will be dealt with later in the following section.

2.2 *Information needs of different users*

A system related to the PIS is a *Pharmaceutical Management Information System (PMIS).* A PMIS is a system that can synthesize large volumes of data generated by pharmaceutical management operations. A PIS should be able to communicate with a PMIS. The PMIS uses data sent by PISs for use in planning activities, estimating demand, allocating resources, and monitoring and evaluating pharmaceutical management operations. (Embrey and for Health (Firm), 2013).

A PMIS should be able to meet the information needs of users with different requirements. In the MDS book, (Embrey and for Health (Firm), 2013) they discussed different information needs

---

1. This was developed by an undergraduate team in the University of the Philippines Manila that the author was a part of.
2. Another option for pharmacies is to just buy a new PIS, but this does not work all the time. In some cases, buying a new PIS may mean more cost and more time because training is needed. Sometimes, a new system is not accepted by the staff since they are more comfortable with the old system and the new system may not connect with the old ecosystem. That is why what is suggested in this paper is improving an existing PIS.
3. http://apps.who.int/medicinedocs/en/m/abstract/Js19577en/

Facility		
Use	Medical director, pharmacist	Prescription patterns
		Patient adherence
		Pharmaceutical availability
		Patient load
Inventory control	Storekeeper	Maximum and minimum stock levels
		Lead times for requisitions
		Prices
		Medicine-use rates
		Shelf life
		Cold-storage temperature variations

Figure 1. Information users and information needs (Embrey and for Health (Firm), 2013).

of various users. Figure 1 shows the information needed from facilities, or in this case, pharmacies; PISs should be able to produce this information to be sent to warehouses and PMISs.

2.3 *RxSolution*

One example of a good pharmaceutical management system is *RxSolution*. RxSolution has been used in more than 200 sites in southern Africa.[4]

RxSolution is an integrated computerized pharmaceutical management system that's used to manage inventory, process purchase orders, handle issues to wards, out/inpatient pharmacy, and satellite clinics, dispense medication to patiens, and prepare repeat prescriptions for down-referral at the facility level. RxSolution supports best practices for procurement, storage, distribution, and dispensing of pharmaceuticals and medical supplies to help always ensure availability of critical products at all times. It has modules for: Budgeting, Procurement, Receipts, Requisitions, Dispensing, and Down-referrals. (Embrey and for Health (Firm), 2013).

One limitation of RxSolution is that it does not have connectivity with other systems right off the box. It still must be manually integrated into an existing financial package to get complete health management information system experience.[5] This limitation of PISs can be solved by implementing interoperability standards. This will be shown in the *connectivity to other systems* subsection under the *methodology section* of this paper.

2.4 *Medication management system*

The specific PIS that will be studied in this paper is named *Medication Management System (MMS)*. This was an undergraduate project by a team of University of the Philippines Manila undergraduate students for a local pharmacy; wherein the author was a part of the team. MMS will serve as an example of how to update a closed-world PIS to an open-world PIS.

3 METHODOLOGY

For this study, due to the limitation of time and resources, the researcher will not create a proposal of improving every aspect of the PIS but only some parts that are essential for it to adapt to the open-world setting. Note that only the Philippine context that will be considered for in this paper, *i.e.*, the researcher will only consider what's applicable to the Philippines.

3.1 *Medication Management System*

As mentioned in the literature review, the PIS that will be used as an example in this paper is the *Medication Management System (MMS)*. MMS has the following main modules:

---

4. http://siapsprogram.org/tools-and-guidance/rxsolution/
5. https://www.msh.org/resources/rxsolution

1. Activity Log
2. Dispense Product
3. Add/Edit Products
4. Order Product
5. Receive/Edit Deliveries
6. Reports

Under the Dispense Product module, it gives the user an option if the dispensing is for inpatient or outpatient. Billing and pricing are done in this module as well, it is processed after dispensing.

Under the Add/Edit Product module, there is a submenu that allows users to choose from the following actions: add new product, edit product, edit vendor, and edit vendor product; each with its own forms.

Under the Reports module, the user can select what type of report they will create. The user can generate a Summary of Issuance Report, Deliveries Report, Inventory Report, or Reorder Summary Report.

3.2 Modules of a PIS

In a primer on PISs by David Troiano (Troiano, 1999), he stated that the following are basic functions commonly needed in a PIS:[6]

- Inpatient order entry, management, and dispensing
- Outpatient order entry, management, and dispensing
- Inventory and purchasing management
- Reporting (utilization, workload, and financial)
- Clinical monitoring - this module allows pharmacists to monitor drug interactions, drug allergies, and other possible medication-related complications
- Manufacturing and compounding - most pharmacies perform some type of compounding, wherein they take commercially available products and change their strength or dosage form; this module supports this activity by providing automated logbooks wherein the pharmacy can record information
- Intervention management - this module is used to document activities and effects of medication therapy.
- Medication administration - this module is used to assist nursing staffs in determining when to administer medication.
- Connectivity to other systems
- Pricing, charging, and billing

Based on this primer, the MMS lacks some of the modules, namely *clinical monitoring, manufacturing and compounding, intervention management, and medication administration.* When developing a PIS, for it to function fully, all of these modules should be included. The MMS shows the case wherein some PIS have been developed with lacking modules, that even the clients (pharmacy owner, doctor) do not know.

This primer should serve as a guide for PIS developers in what modules to include in the PIS. It can be used in the planning or requirements specification phase of software development.

3.3 Connectivity to other systems

Another lacking property of MMS the researcher would like to point out is the *Connectivity to other systems.* The PIS should be able to interact with other systems (*e.g.* a billing system), which means it should be able to send data to other systems and receive data from them as

---

6. Some of the terms have been explained but the other terms are self-explanatory

well. As discussed by *Troiano et al.* (Troiano, 1999), one issue in connectivity of the PIS to other systems is the frequent changing of orders, which requires continual interchange between two systems to make sure that the data is correct.

Another term used synonymously to connectivity to other systems is the word *interoperability*. Interoperability, as defined by the Healthcare Information and Management Systems Society[7], is the ability of different information systems, devices, or applications to connect, in a coordinated manner, within and across organization boundaries to access, exchange and cooperatively use data amongst stakeholders, with the goal of optimizing health of individuals and populations." (Information and (HIMSS), 2019) Thus, to fulfill this need of interoperability, health data exchange architectures and standards could be implemented to allow relevant data to be shared effectively and securely within all applicable settings and relevant stakeholders.

Interoperability can be implemented in 4 different ways: (Information and (HIMSS), 2019)

- "Foundational" - done by establishing the inter-connectivity requirements needed for one system or application to share data with and receive data from another.
- "Structural" - done by defining a structure or format of data exchange (*i.e.*, the message format standards) where there is uniform movement of healthcare data from one system to another such that the purpose and meaning of the data is preserved and unchanged.
- "Semantic" - a hybrid implementation of *foundational* and *structural*.
- "Organizational" - this implementation encompasses the technical components as well as clear policy, social and organizational components (*i.e.*, *non-technical aspects*).

In this study, the researcher will discuss how to implement *structural* interoperability or *interoperability standards* since it is what's available in the Philippine context.

There are many existing interoperability standards that a PIS can implement. One example is *Healthcare Interoperability Resources (FHIR)* Specification by *Health Level Seven (HL7)*.[8] But since we are working in the Philippine context, one important thing to factor in is compliance to the Republic Act 10173 - Data Privacy Act of 2012.[9] Thus, we recommend using the standard developed by the Department of Health of the Philippines called *PHIE Lite*; this interoperability standard is currently being adopted and continuously developed as part of the *Philippine eHealth Strategic Plan*[10] specifically for the *Philippine Health Information Exchange* initiative.[11] So, for a closed-world PIS in the Philippines to update to an open-world setting, the DOH Interoperability standard called *PHIE Lite*[12] must be implemented within the system.

3.4 *Improving the code*

Aside from improving the interoperability of the PIS to other systems and adding the needed modules, the PIS can also be generally improved in terms of performance and function by using software engineering techniques on its source code. These techniques, when implemented correctly, could help the developer add modules easier, thus, making the system more flexible as well - a requirement of a program in the open-world setting.

3.4.1 *Modularization*
One software engineering technique that can be used in PISs is *modularization*. In Parnas' paper, *On the Criteria To Be Used in Decomposing Systems into Modules*, (Parnas, 1972a) he suggested criteria that can be used in decomposing system into modules: (Parnas, 1972a)

---

7. https://www.himss.org/
8. http://hl7.org/fhir/STU3/overview.html
9. https://www.privacy.gov.ph/data-privacy-act/
10. http://ehealth.doh.gov.ph/index.php/pehsp/overview
11. http://ehealth.doh.gov.ph/index.php/phie/overview/40-phie
12. Access for the documentation of PHIE Lite and testing for compliance can be requested at http://phielitet est.doh.gov.ph

1. A data structure, its internal linkings, accessing procedures, and modifying procedures are part of a single module
2. The sequence of instructions necessary to call a given routine and the routine itself are part of the same module.
3. The formats of control blocks used in queues in operating systems and similar programs must be hidden within a "control block module."
4. Character codes, alphabetic orderings, and similar data should be hidden in a module
5. The sequence in which certain items will be processed should be hidden within a single module

By using these criteria, the developer can modularize their system and reap the benefits of modular programming, which are: (Parnas, 1972a)

1. Managerial benefits - development time is shortened because groups can work independently
2. Product Flexibility - it makes change possible to one module without changing the other modules
3. Comprehensibility - it makes understanding the source code easier by studying it one module at a time

Thus, by using modular programming techniques, we can improve PISs in general and make them more flexible for future technologies.

In Parnas's paper, he also noted that a careful job of decomposition can result in easier considerable carryover of work from one project to another. This means that using modularization can also improve the maintenance of the PIS.

A tool called Model-View-Controller (MVC) framework can be used to help implement these principles of modularization. An MVC framework is an architectural pattern commonly used for developing user interfaces that divides an application into three interconnected parts. The MVC design pattern decouples the major components, allowing efficient code reuse and parallel development. (Wikipedia, 2019)

There are many MVC frameworks for different programming languages. Most of them are opensource and are available in the Internet. But since MMS is written in PHP, an MVC framework that can be implemented for this specific case is *CodeIgniter*. CodeIgniter is an MVC framework for web applications written in PHP.[13]

For the case of MMS, the missing modules could easily be added due to the structure of the MVC framework. For each missing module, an additional controller file could just be added with its supporting model and view files. This can be done without modifying much of the existing modules already created.

3.4.2 *Documentation*

As mentioned in the literature review of this paper, a good PIS system needs documentation and a user's manual. (Embrey and for Health (Firm), 2013) MMS currently don't have these documents, hence, it is important to develop these requirements now in an efficient way. Note that best practices dictate that documentation should begin before the coding phase of a project, not after coding. But for this specific case, since MMS has already been developed, and some other PISs might have been developed without documentation. Thus, we will proceed to discuss how good documentation can be produced for existing PISs.

In this section, the creation of a non-technical documentation or a user's manual will not be discussed, since that could be created by non-technical personnel who use the system. Rather, the development of technical documentation will be discussed.

For the documentation on a program's external behavior, there are many techniques that can be implemented: Finite State Machines, Decision Tables and Decision Trees, Program Design Language, Structured Analysis/Real Time, Statecharts, Requirements Engineering

---

13. https://codeigniter.com/

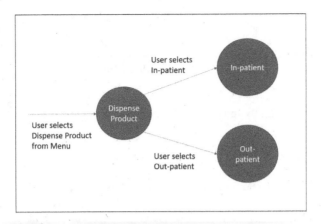

Figure 2.   Using statecharts for dispense product module.

Validation System, Requirements Language Processor, The Specification and Description Language, PAISLey, Petri-nets. (Davis, 1988) All of these techniques won't be discussed in the paper, for further details on these techniques, see Alan Davis's paper (Davis, 1988).

For the specific case of MMS, since PISs are real-time system, *statecharts* could be used. *Statecharts* are extensions to Finite State Machines by Harel. (Harel, 1987) Statecharts make it easier to model complex real-time system behavior without ambiguity. Harel's statecharts provide natural extensions to Finite Sate Machines, making them suitable for specification of the external behavior of real-time systems. An example of using a statechart for the *dispense product* module is seen in Figure 2.

For the internal documentation or source code documentation of the PIS, Parnas's method for software module specification (Parnas, 1972b) could be used. Parnas's method is straightforward for each function, simply specify:

1. The set of possible values or data types
2. Initial values
3. Parameters
4. Effect

Since the PHP language is being used, and PHP uses dynamic typing, there is no need to include the set of possible values or data types. Only the initial values, parameters, and effect of each function. And because the MVC framework is being used, this can be applied to every function inside the controller and model files. It does not need to be applied to view files, a simple description of what the view file displays will suffice.

4   DISCUSSION

The goal of this paper is to show that it is feasible to update a closed-world PIS to an open-world PIS in the Philippine context. As proof of concept, an undergraduate local PIS, MMS, was used as a closed-world PIS example. The goal was achieved by showing methods that can be used to update MMS:

- MMS can be modularized by using an MVC framework which improves the flexibility and writability of the PIS.
- Due to the modularized nature of the code, the documentation of the source code can now be easily written thus improving the maintainability of the system.
- The external behavior of the system could be documented by using *statecharts*. (Harel, 1987)
- Its connectivity to other systems or its interoperability can be improved by implementing DOH's PHIE Lite.

7

After applying these techniques, the missing modules from the list of required PIS modules (Troiano, 1999) could be added more easily. Thus, completing the functionality of the PIS.

By this, we have shown that MMS can be improved to be a PIS that is usable in the open-world environment since it would have been made flexible, maintainable, and interoperable. These methods show that it is feasible even for a local PIS made by undergraduate students to be updated to the open-world. Generally, this can serve as a guide for other local PIS in the Philippines who are looking to update their system.

4.1 Limitations and further research

The methods discussed here are conceptual and are not actually implemented. But the methods discussed are realistic, implementable and available for any developer. For further research, it could be implemented so it can be studied more in detail. Furthermore, a framework for improving a PIS could be developed to formalize the process and make it easier to follow.

5 CONCLUSION

In this paper, the researcher has shown that it is feasible to convert *MMS*, an example of a closed-world PIS, to an open-world PIS in the Philippine context by using modern and classic software engineering techniques available today. This paper can serve as a guide for Philippine PISs that want to convert to an open-world setting.

REFERENCES

Alanazi, A., Rabiah, F. A., Gadi, H., Househ, M., and Dosari, B. A. (2018). Factors influencing pharmacists' intentions to use pharmacy information systems. *Informatics in Medicine Unlocked*, 11:1–8.

Baresi, L., Di Nitto, E., and Ghezzi, C. (2006). Toward open-world software: Issues and challenges. *Computer*, 39(10):36–43.

Chapuis, C., Roustit, M., Bal, G., Schwebel, C., Pansu, P., Tchouda, S., Foroni, L., Calop, J., Timsit, J.-F., Allenet, B., jean luc, B., and Bedouch, P. (2010). Automated drug dispensing system reduces medication errors in an intensive care setting. *Critical care medicine*, 38:2275–2281.

Davis, A. M. (1988). A comparison of techniques for the specification of external system behavior. *Commun. ACM*, 31(9):1098–1115.

Embrey, M. and for Health (Firm), M. S. (2013). *MDS-3: Managing Access to Medicines and Health Technologies*. MDS-3: Managing Access to Medicines and Health Technologies. Kumarian Press.

Harel, D. (1987). Statecharts: A visual formalism for complex systems. *Sci. Comput. Program.*, 8(3):231–274.

Information, H. and (HIMSS), M. S. S. (2019). What is interoperability? Accessed: 2019- 04-10.

Kazemi, A., Rabiei, R., Moghaddasi, H., and Deimazar, G. (2016). Pharmacy information systems in teaching hospitals: A multi-dimensional evaluation study. *Healthcare Informatics Research*, 22:231.

Parnas, D. L. (1972a). On the criteria to be used in decomposing systems into modules. *Commun. ACM*, 15(12):1053–1058.

Parnas, D. L. (1972b). A technique for software module specification with examples. *Commun. ACM*, 15(5):330–336.

Troiano, D. (1999). A primer on pharmacy information systems. *JOURNAL OF HEALTHCARE INFORMATION MANAGEMENT*, 13(3):41–52.

Wikipedia (2019). Model-view-controller. Accessed: 2019-04-18.

Theory and Practice of Computation – Nishizaki et al (eds)
© 2021 Taylor & Francis Group, London, ISBN 978-0-367-41473-3

Comparison of the performance of MLP, random forest, and adaboosted random forests in software cost estimation

I.K.P. Paderes, T.M. Mendez & L.L. Figueroa
University of the Philippines, Diliman, Philippines

ABSTRACT: Even with the eve of state-of-the-art development tools and new development techniques, there is still no standard to a proper software effort estimation that can accurately pre-compute the amount of man-hours of a complete software project. In this study, three machine learning models (Multilayer Perceptrons, Random Forest, and AdaBoosted Random Forest) were explored to compute the software development effort using COCOMO values. This study used aggregated datasets taken from the PROMISE Repository to train the learning models and these models were applied to data of different software development companies. Results show that the AdaBoosted Random Forest Model gives out the best performance among the three and can provide a modern substitute to software effort estimation.

Keywords: software cost estimation model, COCOMO, artificial neural networks, ensemble learners

1 INTRODUCTION

Project cost estimation is the process of predicting the quantity, cost, and price of the resources required by the scope of a project. In Software Engineering, the length of project development influences the overall cost of a software product to a great extent. Effort estimation, along with other factors such as development time estimation, cost estimation, team size estimation, risk analysis, etc. are calculated in the early phases of a project. A good model for calculating these parameters will help prevent potential loss for the business as software development can be very expensive.

The Constructive Cost Model (Boehm, 1981) developed by Barry Boehm in the late 1970s has been able to help people understand the cost consequences of the decisions they will make in commissioning, developing, and supporting a software product. Besides providing a software cost estimation capability, COCOMO provides a great deal of material which explains exactly what costs the model is estimating, and why it comes up with the estimate it does. (Boehm, 1984).

In the paper of Goyal and Parashar (2018), they claimed that over the years, development methods have undergone revolutionaries but estimation techniques seems to be lagging as they are not so modified to cope up with the modern development skills. Goyal and Parashar stressed out the need of training models to work with updated development methods. The COCOMO model has been widely used for software effort estimation but its accuracy could be improved by applying modern techniques.

Machine Learning has become a key technique for solving problems in areas such as Computational Finance, Image Processing, Computer Vision, Computational Biology, Energy Production, Automotive, Aerospace, Manufacturing, and Natural Language Processing. Its many applications inspired a few researchers to explore its use in solving the cost estimation problem.

Goyal and Parashar (2018), Reddy and Raju (2009), and Kaushik et al. (2012) created Multilayer Perceptron Models (MLP) to compete with the performance of the COCOMO model. Bhatia and Attri (2015)implemented a Decision Tree for effort estimation. Nassif et al.

(2013) also explored the application of Decision Trees and Random Forest Models (RF) for effort estimation.

In this study, the researchers aimed to explore other machine learning models that can give better estimation of software engineering effort. Specifically, they aimed to improve the performance of Random Forest model by applying AdaBoost algorithm. Later, they compared this model with MLP and RF.

2 BACKGROUND

Over the years there have been an effort in exploring the capacity of artificial intelligence in predicting the effort (man-hours) required to successfully complete a project to avoid the consequences of erroneous budget. Researchers apply machine learning techniques on top of COCOMO to improve the accuracy of software cost estimation.

2.1 *COCOMO*

The Constructive Cost Model (COCOMO) is a Software Estimation Model that was developed by Barry Boehm in the late 1970s during his tenure with the Defense Systems Group at the TRW Inc., and published in 1981 in his book "Software Engineering Economics". This model is a hierarchy of three models: Basic, Intermediate, and Detailed Model. It is based on a study of 63 projects developed at TRW from the period of 1964 to 1979. (Boehm, 1981).

Table 1 . Coefficients for Intermediate COCOMO.

Mode	a	b
Organic	3.2	1.05
Semi-detached	3	1.12
Embedded	2.8	1.20

The Basic Model estimates effort for the small to medium sized software projects and takes the form:

$$E = a(SIZE)^b \tag{1}$$

where E is effort applied in person-months and SIZE is measured in thousand delivered source instructions. In standard COCOMO, there are exactly 152 hours per person-month. The basic terms of measurement for this model is the number of lines of code, usually computed by the thousands which is denoted by KLOC. The coefficients a and b are dependent upon the following modes of development of projects:

(a) Organic mode - for small-sized projects between 2-50 KLOC with experienced developers in a familiar environment
(b) Semi-detached mode - for medium-sized projects between 50-300 KLOC with developers with average previous experience on similar projects
(c) Embedded mode - for large and complex projects typically over 300 KLOC with developers having very little previous experience

The Intermediate Model takes into account the software development environment. Boehm introduced a set of 15 cost drivers in the Intermediate COCOMO that adds accuracy to the Basic COCOMO. The cost drivers are grouped into four categories:

1. Product attributes
   (a) Required software reliability (RELY)
   (b) Database size (DATA)
   (c) Product complexity (CPLX)

2. Computer attributes
   (a) Execution time constraint (TIME)
   (b) Main storage constraint (STOR)
   (c) Virtual machine volatility (VIRT)
   (d) Computer turnaround time (TURN)

3. Personnel attributes
   (a) Analyst capability (ACAP)
   (b) Application experience (AEXP)
   (c) Programmer capability (PCAP)
   (d) Virtual machine experience (VEXP)
   (e) Programming language experience (LEXP)

4. Project attributes
   (a) Modern programming practices (MODP)
   (b) Use of software tools (TOOLS)
   (c) Required development schedule (SCED)

The cost drivers have up to six levels of rating: Very Low, Low, Nominal, High, Very High, and Extra High. Each rating has a corresponding real number known as effort multiplier, based upon the factor and the degree to which the factor can influence productivity. The estimated effort in person-months (PM) for the intermediate COCOMO is given as:

$$Effort = a \times [SIZE]^b \times \prod_{i=1}^{15} EM_i \qquad (2)$$

In equation (2), the coefficient a is known as the productivity coefficient and the coefficient b is the scale factor. They are based on the different modes of project as given in Table 0. The contribution of effort multipliers corresponding to the respective cost drivers is introduced in the effort estimation formula by multiplying them together. The numerical value of the ith cost driver is EM_i (Effort Multiplier).

In their paper, Kaushik et al. (2012) used the intermediate COCOMO model because it has estimation accuracy that is greater than the basic version, and at the same time comparable to the detailed version. In this study, the researchers will also be using the intermediate COCOMO model.

2.2 *Machine learning models*

Software estimation models are categorized into Algorithmic and Non-algorithmic techniques. Algorithmic models are formula-based models derived from project data. These compute effort by performing some calibration on the pre-specified formulae. (Goyal & Parashar, 2018) While the COCOMO model is classified under algorithmic models, Machine Learning-based effort estimation is a non-algorithmic approach.

Effort estimation can be classified as a regression problem (i.e. predicting a continuous value given features of the software that you want to build). Random Forest, Neural Network, Gradient Boosting Tree, Decision Tree, and Linear Regression are the most appropriate machine learning models to use when solving regression problems.

The field of artificial neural networks is often just called neural networks or multilayer perceptrons (MLP). The power of neural networks come from their ability to learn representations in a training data and how to best relate it to the output variable that you want to predict. Mathematically, they are capable of learning any mapping function and have been proven to be a universal approximation algorithm.

A Decision Tree (DT) is a logical model represented by a binary tree that illustrates how a target variable (a.k.a. dependent variable in regression models) is predicted using a set of predictor variables. The main advantage of a DT model is that it can help non-technical people to see the big picture of a certain problem. However, DT models might suffer from the overfitting problem, as well as from providing good accuracy in comparison to other models. For their paper, Nassif et al. (2013) used the DTREG program (Sherrod, 2011) to generate a DT model for effort estimation.

Random Forest (RF) is an ensemble learning method for classification and regression problems. It is an unsupervised learning method that builds multiple decision trees at training time and outputs the class (i.e. mode of the classes for classification or mean prediction for regression) of the individual trees.

Boosting is an ensemble technique that attempts to create a strong classifier from a number of weak classifiers. AdaBoost, short for "Adaptive Boosting", is the first practical boosting algorithm proposed by Freud and Schapire in 1996.

3 RELATED WORK

This section presents publications on effort estimation models that utilize machine learning methods.

Reddy and Raju (2009) constructed a software effort estimation model based on artificial neural networks. The model was designed to improve the estimation accuracy of COCOMO model. They proposed a multilayer feedforward neural network to accomodate the model and the parameters to estimate software development effort. The network was trained with back propagation algorithm by iteratively processing a set of training samples and comparing the network's prediction with the actual effort. The results obtained suggested that the proposed architecture can be replicated for accurately forecasting software development effort.

Kaushik et al. (2012) used artificial neural networks to model the COCOMO estimation model. The neural network that they used to predict the software development effort was trained using the perceptron learning algorithm. They used the COCOMO'81 dataset to train and test the network and found that the obtained accuracy of the network is acceptable.

Goyal and Parashar (2018) constructed machine learning models trained using the COCOMO dataset and tested using the KEMERER dataset. Their best model, SG5, showed a Mean Magnitude of Relative Error (MMRE) of 228.70%. They claimed this to be superior to the reported accuracy of COCOMO (i.e. 284.61%). Their models, SG1 through SG6 are fairly accurate and strongly correlated. Their results shows that machine learning is an exciting new approach to the cost estimation problem.

Artificial Neural Network is not the only machine learning model applied to the cost estimation problem. Nassif et al. (2013) conducted a study that compares the performance of Decision Trees and Decision Tree Forest Models on this particular problem. Their models were developed using 10-fold cross-validation technique using the ISBSG and Desharnais datasets. Results of this study have shown that the DTF model is competitive and can be used as an alternative in software effort prediction.

In this study, the researchers will explore three machine learning models: Multilayer Perceptron Model, Random Forest Model, and AdaBoosted Random Forest Model.

4 METHODOLOGY

4.1 *Experimental setup*

The proposed methodology for the comparison of all of the models' performance in estimating the development effort required by software projects can be visualized in Figure 1

Step I. Define the datasets to be used for training and testing with sets for the input vectors and the output vector. This includes processing the inputs for optimal training of the model.

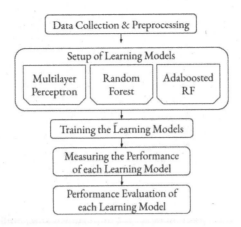

Figure 1.   Experimental flow diagram.

The input dataset consists of independent attributes and the output values is comprised of the actual dependent effort attribute.

Step II. Implement the networks of the three learning models to be compared. All of the hyperparameters to be used by each network should be setup and applied. Then, the networks are initialized for training.

Step III. For each learning model, the training dataset processed in *step I* is used to feed the learning model's network to create the associations between the data which are going to be used further in effort estimation.

Step IV. After training the networks, their performance are measured by using the testing dataset from *step I*.

Step V. The networks' performances are then compared with each other.

4.2   *Data collection and preprocessing*

4.2.1   *Data collection*
For any learning model to be effective in estimating development effort, there must be enough data to train on and associate with. To help train the models, public COCOMO datasets from the PROMISE Software Engineering Repository (Sayyad Shirabad & Menzies, 2005) were compiled into one. The PROMISE repository contains COCOMO values from different software projects (Barry Boehm's Software Engineering Economics and NASA projects from 1971s-1990s).

To test the performance of the models, the researchers made use of data from Kemerer's dataset (1987). This dataset were compiled from projects of a national computer consulting and services firm. The use of these projects as a testing dataset were already justified by Kemerer 1987 as their projects are well-documented and their output is of consistent professional quality.

After testing the performance of the models, the researchers then tried to predict data from a software project developed by a local software development company (hereinafter referred in this paper as ABC company). This paper followed the COCOMO values described by Boehm where the project's COCOMO cost drivers were assessed as influenced by product attributes, computer attributes, personnel attributes, and project attributes. Each of these cost drivers was rated either Very Low, Low, Nominal, High, Very High, or Extra High and was later converted to their corresponding numerical values.

4.2.2 *Data preprocessing*
After the compilation of training and testing datasets, minor preprocessing must still be applied as the performance of learning models are very sensitive to data. To address this issue, the dataset column values were transformed and reduced. For the transformations, first the KLOC and the actual effort values were transformed into their corresponding logarithmic values. Next, the mode of the projects were transformed into a three-digit boolean value

where Embedded = 1 0 0, Semi-detached = 0 1 0, and Organic = 0 0 1. These transformations were derived from the preprocessing done by Goyal and Parashar. These transformations decreases the range between the minimum and the maximum input variable.

To reduce the sensitivity of learning models to the inputs further, the transformed dataset had its features reduced next by taking only its top features. To do this, the researchers took advantage of the built-in variable importance of the random forest model which is the Gini Importance. Gini Importance calculates the total decrease of node impurity averaged over all trees of the ensemble and is given as

$$Gini\ Importance = 1 - \sum_i P_i \qquad (3)$$

where P_i is defined as the proportion of subsamples belonging to a certain target class. To help choose the features, the researchers set an importance threshold of 0.01 to remove features that have lower Gini Importance value than the threshold. Only the values of the effort multipliers and the KLOC were ranked as the mode values were used to classify each project into their respective modes. With the Gini Importance function, the researchers were able to reduce the features of the input dataset from 18 to only 10. Information about the final training and testing datasets is described further in Table 2.

4.3 *Proposed setup of the learning models*

Each of the learning models will have a different network architecture from each other except for their input and output layers. Given that all of these models are regressors, they will have ten nodes (the number of the final input features) for their input layer and one node for the output layer.

For the multilayer perceptron, the researchers used the hidden layer and node configuration of the best model formulated by goyal. This configuration uses two hidden layers, the first one having 18 neurons and the second one having 15. For the activation and solver functions, the researchers used the default functions offered by the ***MLPRegressor*** function of Scikit-learn. A rectified linear unit function was used as the activation function between the input and the hidden layers while an identity activation function was used between the last hidden layer to the output layer. For its solver function for weight optimization, a stochastic gradient-based optimizer called *Adam* (Kingma & Ba, 2014) was used. A visualization of the MLP setup is seen in Figure 2.

Table 2.   Transformed training and testing dataset information.

Dataset	Shape	Input Features	Output Features
Training	216 x 10	log(KLOC), TIME, STOR, DATA, RELY, MODP, CPLX, MODEA, MODEB, MODEC	log(EFFORT)
Testing	15 x 10	log(KLOC), TIME, STOR, DATA, RELY, MODP, CPLX, MODEA, MODEB, MODEC	log(EFFORT)
ABC Company Project	1 x 10	log(KLOC), TIME, STOR, DATA, RELY, MODP, CPLX, MODEA, MODEB, MODEC	log(EFFORT)

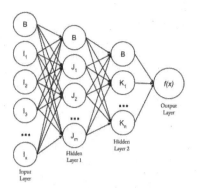

Figure 2.   Multilayer perceptron setup.

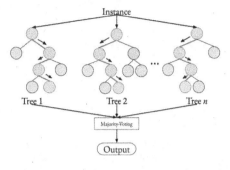

Figure 3.   Random forest setup.

To have a better range of comparison for the AdaBoosted Random Forest regressor, the researchers added Random Forest as one of the learning models to compare. It uses the configuration formulated by Nassif which has a maximum of 500 decision trees, a minimum size node split condition of 2, and a maximum depth level of 100. Figure 3 shows how Random Forests generate their output from the ensemble of decision trees.

Lastly, as the AdaBoosted RF learning model is an ensemble learning model, it will have weak Random Forest regressors learning inside its network. The main ensemble model will train 300 random forests and the configuration used for each weak RF learners has only 100 decision trees and a maximum depth level of 2. A sample visualization on how the AdaBoost algorithm builds its model is seen in Figure 4.

4.4   *Performance evaluation criteria*

The performance of each models can be evaluated by the degree in which its estimated output matches the actual effort of the project itself. Any estimation has a problem of being an overestimation or an underestimation. Overestimation may lead into less production or having suffered the "gold plating" phenomenon. (Boehm, 1981) Underestimation may lead to the project being understaffed and might even make the project suffer Brooke's law (Boehm, 1995) where managers adding more man-power to the project with the hopes of finishing it faster will make it finish later. Therefore, the criterion to evaluate the learning model's performance must take these factors into consideration.

boehm introduced the percentage error test given as:

$$Percentage\ Error = \frac{Effort_{Estimated} - Effort_{Actual}}{Effort_{Actual}} \tag{4}$$

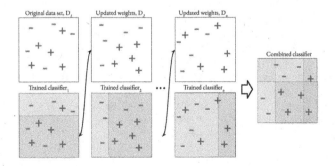

Figure 4.   AdaBoosted RF setup.

15

but this equation may give out estimation varying from over-estimation (negative percentage error) or under-estimation (positive percentage error) Conte, Dunsmore, and Shen suggested a magnitude of relative error (MRE) test which takes under consideration both under-estimation and over-estimation. MRE is given as:

$$MRE = \left| \frac{\tilde{y}_i - y_i}{y_i} \right| \qquad (5)$$

where y_i is the actual effort value while $\tilde{y}_i$ is the estimated effort value. With the MRE as the base calculation for accuracy for each estimate, a performance evaluation criteria over all of the testing dataset is given by the MMRE. Mean MRE is given as:

$$MMRE = \frac{\sum_{i=1}^{i=n} MRE_i}{n} \qquad (6)$$

where n is the number of estimates, and MRE_i is the MRE of the *ith* estimate. The Standard Deviation of the MRE (SDMRE) and the Median MRE (MdMRE) for all n estimations can also be adopted as performance evaluation criteria. For all of these criteria, the perfect test result would be zero.

Another criterion to be considered is the Correlation Co-efficient (R^2). This reflects the degree in which the model's estimation correlate with the actual results. The perfect value of the R^2 would be one.

Lastly, to complement MMRE, PRED(x) can also be computed to check the predictive power of the model. It computes the ratio of the regressor in which its MRE is less than or equal to x over the number of estimates. It is given as:

$$PRED(x) = \frac{k}{n} \qquad (7)$$

where k is the number of projects where $MRE_i \leq x$ and n is the total number of observations. The researchers assigned the value of 0.25 to be x.

4.5 *Program description*

The program used for the study consists of two parts, the preprocessing functions and the models themselves. The program is developed mainly using Python with the usage of the Scikit-learn machine learning library for the creation of the learning models.

5 RESULTS AND DISCUSSION

For the training and testing datasets, the features to be used for learning was reduced. This was done with the use of their Gini-Importance ranking. With the limit threshold of 0.01, the top features of the datasets are found in Table 1.

This study aimed to come up with machine learning models that can give an accurate effort estimation for software engineering projects. In addition, the researchers aimed to compare the performance of these models with respect to MMRE, MdRE, SDMRE, R^2 and PRED.

After setting up the training and testing datasets, three types of machine learning models were then constructed in this study: (1) Multilayer Perceptron (MLP); (2) Random Forest (RF); and (3) AdaBoosted Random Forest (ADF).

Table 3 .   Gini-importance ranking.

Input Feature	Gini-Importance (> 0.01)
KLOC	0.8077910314
TIME	0.0575454769
STOR	0.01742975346
RELY	0.01620340017
LEXP	0.01532081361
CPLX	0.01460867421
MODP	0.01429471377

Figure 5 Compares the actual effort values with the newly introduced Adaboosted RF model's and the COCOMO model's results. Figures 6, 9, and 12 compare the actual development effort value to the prediction of each learning model. Shown in Figures 7, 10, and 13 are the correlation of the predictions to the actual effort values. Figures 8, 11, and 14 show the MRE of the learning models for each project inside the testing dataset. An overview of the performance metrics measured from the learning models is shown in Table 4.

Among the three models, the AdaBoosted Model has shown the least Mean Magnitude of Relative Error (MMRE) at 29.5%. This is relatively better than the MMRE of MLP and RF at 31% and 30.8% respectively.

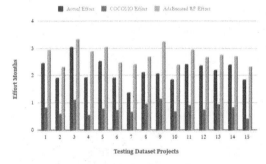

Figure 5.   Actual effort vs COCOMO effort vs adaBoosted RF effort.

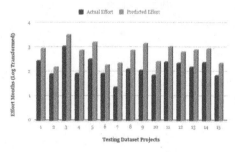

Figure 6.   Actual vs predicted bar graph of multilayer perceptron.

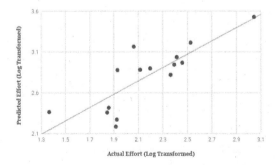

Figure 7.   Actual vs predicted scatter graph of multilayer perceptron.

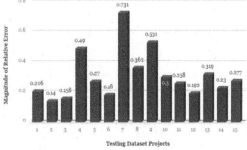

Figure 8.   MRE of the multilayer perceptron model to the test dataset.

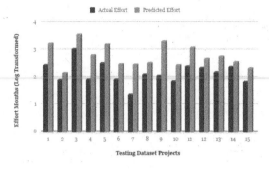

Figure 9. Actual vs predicted bar graph of random forest.

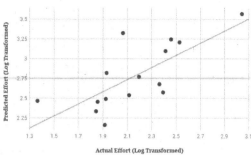

Figure 10. Actual vs predicted scatter graph of random forest.

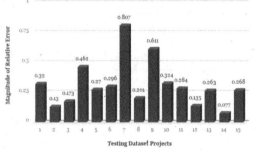

Figure 11. MRE of the random forest model to the test dataset.

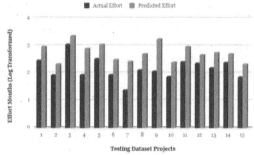

Figure 12. Actual vs predicted bar graph of ada-boosted random forest.

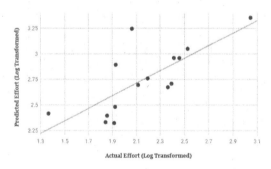

Figure 13. Actual vs predicted scatter graph of ada-boosted random forest.

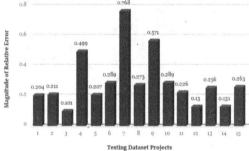

Figure 14. MRE of the adaBoosted RF model to the test dataset.

Table 4. Performance overview of the learning models.

	MLP	RF	AdaBoosted RF
MMRE	**0.31**	0.308	0.295
SDMRE	0.156	0.185	0.176
MdMRE	0.2704	0.2697	0.256
PRED(0.25)	0.40	0.333	**0.467**
R^2	**0.818**	0.729	0.74

Table 5. MRE of ABC company project on all learning models.

	ABC Company Project MRE
Multilayer Perceptron	2.8586
Random Forest	2.6927
AdaBoosted RF	**2.2789**

Another criterion considered is the Correlation Coefficient (R^2). The closer the correlation coefficient is to 1, the better. MLP showed a correlation coefficient of 0.818 which is significantly higher than 0.729(RF) and 0.74(ADF).

Finally, PRED(x) was also calculated to check the predictive power of the model. It computes the ratio of the regressor in which its MRE is less than or equal to x over the number of estimates. A higher PRED(x) indicates better performance. Among the three models, the AdaBoosted Random Forest has shown the highest PRED(x) (0.467), defeating RF and MLP at 0.333 and 0.4 respectively.

Aside from the Kemerer dataset, the models were also evaluated over a project in ABC Company. Table 5 shows the results of MRE over the three models. Least MRE has been found to be that of AdaBoosted Random Forest at 227%. This is significantly less than the MRE computed for Multilayer Perceptron (285%) and Random Forest (269%).

As seen from the results, even with reducing the parameters the models have shown signs of overfitting to the past historical data. This means that the model has increased estimation accuracy on projects similar to the ones inside its database but not with newer ones like the new project from ABC Company.

Overall, these results have favored the AdaBoosted Random Forest model in solving the effort estimation problem.

6 CONCLUSION

In the software development industry, the accuracy of the effort estimation can bring success or failure to the total development of any software project. Various research have pushed machine learning as a solution to modernize the COCOMO model. Through the presented work, performance among machine learning models was compared in estimating development effort with the use of the COCOMO model. Presented also was the reduction of COCOMO cost driver features with the use of Gini Importance as machine learning regression models were sensitive to input data. The performance evaluation criteria for this study have shown that the AdaBoosted Random Forest model has the best performance in terms of MMRE and PRED(x). Its (R2) value of 0.74 depicts good correlation between the predicted and the actual estimation results. Applied to a modern software project, AdaBoosted Random Forest still has the best performance among the three.

At its current stage, the learning model can be applied within an organization at the earliest stages of a project and data that will be taken from this can also be used to re-train the whole model, therefore producing a model that outputs estimations based on the organizations previous projects. But in order to improve the models performance, it is worth considering the exploration of resampling and validation techniques such as k-fold cross validation or Synthetic Minority Over-sampling Technique in order to minimize oversampling into the dataset. For future work, a working service application can be developed that trains data taken from a database of previous software projects and is continuously updated with new projects.

REFERENCES

Anupama Kaushik, D. M., Ashish Chauhan, & Gupta, S. (2012). Cocomo estimates using neural networks. *I.J. Intelligent Systems and Applications*, 9:22–28.

Bhatia, S. & Attri, V. K. (2015). Implementing decision tree for software development effort estimation of software project. *International Journal of Innovative Research in Computer and Communication Engineering*, 3.

Boehm, B. W. (1981). *Software engineering economics*. Englewood Cliffs, NJ: Prentice-Hall Inc.

Boehm, B. W. (1984). Software engineering economics. *IEEE Transactions on Software Engineering, SE-10*.

Brooks, F. P., Jr. (1995). The mythical man-month (anniversary ed.). Boston, MA, USA: Addison-Wesley Longman Publishing Co., Inc.

Conte, S. D., Dunsmore, H. E., & Shen, V. Y. (1986). *Software engineering metrics and models.*Redwood City, CA, USA: Benjamin-Cummings Publishing Co., Inc.

Goyal, S. and Parashar, A. (2018). Machine learning application to improve cocomo model using neural networks. *International Journal of Information Technology and Computer Science(IJITCS)*, *10*:35–51.

Kemerer, C. F. (1987). An empirical validation of software cost estimation models. *Communications of the ACM, 30*. Retrieved from http://gen.lib.rus.ec/scimag/index.php?s=10.1145/22899.22906 doi: 10.1145/22899.22906

Kingma, D. P. and Ba, J. (2014). Adam: A method for stochastic optimization. *arXiv preprint arXiv:1412.6980*.

Nassif, A. B., Azzeh, M., Capretz, L. F., and Ho, D. (2013). A comparison between decision trees and decision tree forest models for software development effort estimation. In *2013 Third International Conference on Communications and Information Technology (iccit)* (p. 220–224). doi: 10.1109/ICCITechnology.2013.6579553

Reddy, C. and Raju, K. (2009). A concise neural network model for estimating software effort. *International Journal of Recent Trends in Engineering*, 1.

Sayyad Shirabad, J. & Menzies, T. (2005). *The PROMISE Repository of Software Engineering Databases.* School of Information Technology and Engineering, University of Ottawa, Canada. Retrieved from http://promise.site.uottawa.ca/SERepository

Sherrod, P. (2011). DTREG. *Software for Predictive Modeling and Forecasting.*

Theory and Practice of Computation – Nishizaki et al (eds)
© *2021 Taylor & Francis Group, London, ISBN 978-0-367-41473-3*

A data-driven approach to web development

N.M. & L.L. Figueroa
Department of Computer Science, University of the Philippines Diliman, Quezon City, Philippines

ABSTRACT: In this study, we investigated a web development practice that implements data-driven design. Results show that data-driven development is an iterative process integrating web analytics using Google Analytics and Google Optimize as well as heuristic and usability analyses as experience-based assessments. The expertise of the web analyst is crucial to experience-based assessments, which consider relevance, trust, orientation, stimulance, security, convenience, confirmation, and web usability as standards. We evaluated the performance of a website after being improved using the data-driven approach. The case studies reveal that the improved website performs better than the original. However, good traffic to the website, ample time to run experiments, and availability of a live website are necessary for the approach to be applicable. Future work may explore other web analytics tools, employ qualitative research in data gathering, evaluate the performance of other websites, examine other data-driven approaches, and derive a general data-driven web development methodology.

1 INTRODUCTION

One of the challenges in web development is to develop a website that stands out among the millions of websites available in the world and captures its target market. With a target market of more than 4 billion Internet users ("Internet world stats," 2019), there is an infinite opportunity for trading yet there is also an ever increasing challenge of developing a website that can compete with the over 142 million dot-com websites in the world ("Domain count statistics," n.d.; Flavian et al. 2009).

Researches have been made in order to determine the factors that lead to the success of a website in achieving its goal. Flavian et al. (2009) found out that web design, which is one of the stages in web development (Achebe, 2002; Abdul-Aziz et al. 2012; Taylor et al. 2002), is an essential factor for a website to be successful. However, it is not possible to propose a unique optimal design since there are different types of product, different types of user visiting a website, and a website is visited in different geographic locations. Different combinations of design factors are necessary in order to create an optimal website design. Flavian et al. (2009) further emphasized that the user's point of view is important in every aspect of a website. Hence, there is a need to learn more about the users. Liikkanen (2017) introduced data-driven design which involves different tools that can be used to automatically collect and analyze data about users. The analysis can then be used in design research which hopes to produce a website design that leads to a high level of success.

In this study, we investigate an actual web development practice that implements data-driven design. Particularly, we seek to identify the methodology and standards used for data-driven web development. Also, we evaluate how data-driven design affects the performance of a website in terms of achieving its goals.

Web development methodologies had been the subject of study in the past since there had been uncertainty in the way web development was carried out. Taylor et al. (2002) conducted a survey of web development practices among 25 UK organizations. They found out that not all of the 25 organizations used formalized website design techniques, website layout standards, website documentation, and website testing procedures.

Achebe (2002) identified that websites were developed using ad hoc approaches. Institutions developing websites relied on the knowledge and experience of individual developers and on their own development practices. Although there had already been a lot of methodologies, there was still no *de facto* standards. So, Achebe (2002) intended to conduct a comparative study among different web development methodologies from various reputable companies in his time in order to find out concepts applied in web development. The Web Site Design Method (WSDM), an audience-driven approach that focuses on its target audience, was used as the basis of comparing the different methodologies: IconMatrix, Frontline Solutions, and the Reference. WSDM is composed of five main phases: mission statement specification, audience modeling, conceptual design, implementation design, and actual implementation. Except for a few differences with WSDM, all the other methodologies are almost similar in their approach and have proven useful and promising as illustrated by designing a website using the different methodologies. Also, involving the target audience is proven beneficial as it enhances design and development.

Another study was conducted in order to come up with a development methodology for building web-based information systems. The new general methodology was named Web-based Development Life Cycle (WDLC), which was created based on the traditional software development life cycle approach to systems analysis and design. Abdul-Aziz et al. (2012) comprehensively reviewed 14 various web development methodologies and then identified six stages, namely: planning, analysis, design, implementation, publishing, and maintenance, which comprised WDLC. In order to find out the perceptions of web development teams on the WDLC methodology, seven organizations with a common business objective of providing consultancy services in web-based information systems development yet with different web development methodologies were selected as case studies. The authors then interviewed the web development team in each organization in order to gain insights and information regarding the usage of WDLC methodology in developing web-based information systems. By comparing the general WDLC with the development methodologies employed in each participating organization, the authors were able to derive the comprehensive WDLC with five stages, namely: web development planning, web requirements engineering, web application design, web technical implementation, and web system operations. All web development teams of the seven participating organizations agreed with the comprehensive WDLC methodology.

Since there are already millions of websites available nowadays, web development methodologies are challenged to produce a website that captures its target users in spite of the presence of million other websites. Flavian et al. (2009) conducted a study which identified web design, one of the stages in web development, as a key factor for a website to be successful. Although, it is not possible to propose a unique optimal design that will work for all websites, they were able to determine the key aspects for achieving a high level of success of e-commerce websites. The key aspects are the aesthetic appearance of the website, navigation through the website, management of information and content of the website, and the characteristics of the shopping process. In all these aspects, the user's point of view must be emphasized. However, there are many possible users of a website and factors such as Internet user experience, product class knowledge, and cultural differences on users may affect the perceptions about design.

Since many Internet-based services have been available for millions of people, it has become possible to gather information about users and so there is an increasing opportunity to learn more about them. Liikkanen (2017) presented the data-driven approach to web design. Data-driven design is about using tools that automatically collect and analyze data about users. These tools are categorized according to their usefulness for design research: screen recordings, heat maps, experiment management, and descriptive behavioral analytics. Google Analytics is an example product for behavioral analytics while Google Optimize is for experiment management. Other web analytics tools are Adobe Analytics ("Data can come," 2019), Kissmetrics ("Advanced product analytics," n.d.), Mixpanel ("Grow your business," 2019), and Heap ("The new standard," 2019). Our study will investigate how Google Analytics and Google Optimize are used in web development that employs data-driven design.

3 METHODOLOGY

Since our study aims to identify the methodology and standards employed in data-driven web development, we chose the methodology used by Agno Digital Marketing company ("Make more money," n.d.) as our case study. The company's interest is on improving an existing live website such that it gets more visitors take a specific desired action. The goal is to optimize a website so that the business it represents will grow. The company's development team is composed of a web analyst, a web designer, and a web developer. We interviewed the web analyst in order to identify the methodology and standards the company employed as it implemented data-driven design.

We used one of the company's projects called the "Simon Blow Qigong" (2017) website as our case study in showing how data-driven design affects the performance of a website. "Simon Blow Qigong" (2017) is an e-commerce website that promotes a healing system through classes, workshops, retreats, CDs, DVDs, e-books, and downloadable products. The website's main key performance indicators (KPI) are to get more people book for workshops and retreats, get more people buy its products and get more people sign up for membership. We examined some sample reports from the web analyst which showed how the methodology worked and how the website performed before and after data-driven design was applied. In order to verify the correctness of the quantitative data, we had been granted access to the company's Google Analytics and Google Optimize accounts.

4 RESULTS AND DISCUSSION

4.1 *Development methodology and standards*

In this section, we present the methodology and standards employed by Agno Digital Marketing company ("Make more money," n.d.).

The company optimizes a website page by page. The methodology presented below is applied to each page on the website:

(1) Identify the objective of the web page and determine the primary Key Performance Indicator (KPI) and other metrics to improve.
(2) Ask additional questions related to the primary KPI.
(3) Set up the Google Analytics for data collection of all actions a user can do on the web page and track primary KPI.
(4) Gather data and analyze. The activities that can be done are categorized as follows:

 • Experience-based assessment: site walkthroughs, heuristic analysis, usability analysis
 • Qualitative research: online surveys with recent customers, on-site polls, phone interviews, live chat transcripts, customer support insight, user testing
 • Quantitative research: web analytics analysis

This step looks for opportunities disguised as issues on the web page.

(1) List all the findings in step 4 as issues and confirm them through the analytics data.
(2) Translate the data gathered from step 5 into insights and formulate hypotheses based on the insights.
(3) Prioritize each hypothesis by either impact or ease of implementation.

(4) Based on the hypothesis, create design prototype(s) for client's approval to run experiment.

(5) Implement the approved prototype(s) and prepare for experiment.

(6) Run an experiment for a minimum of 7-14 days to determine if the new page is a probable improvement to the original page. The experiment should satisfy these factors: sample size and probability to beat baseline (95%).

(7) Run post-test analysis to determine whether the hypothesis is effective or not. If the hypothesis is effective, replace the original page with the new one.

(8) Prepare reports.

(9) Go back to step 8 while there are hypotheses to test.

In the case of "Simon Blow Qigong" (2017) website, the web analyst conducted heuristic analysis and usability analysis. In conducting heuristic analysis, the web analyst evaluated a web page against the following criteria:

- Relevance. Does my perception fit my expectations?
- Trust. Can I trust this provider?
- Orientation. Where should I click? What do I have to do?
- Stimulance. Why should I do it right here and right now?
- Security. Is it secure here? What if. . .?
- Convenience. How complicated will it be?
- Confirmation. Did I do the right thing?

Usability analysis, on the other hand, was done using the web usability guidelines (Nielsen, 2001; Travis, 2016):

- Homepage usability
- Task orientation
- Navigation and information architecture
- Form and data entry
- Trust and credibility
- Writing and content quality
- Page layout and visual design
- Search usability
- Help, feedback and error tolerance

The outcome of heuristics and usability analyses is what we call "areas of interest" that are validated by the results of web analytics analysis.

Web analytics analysis was done by using Google Analytics and Google Optimize. Google Analytics was used to measure the following metrics:

- Sessions – number of users coming to a website
- Bounce Rate – the percentage of single-page sessions on a website, i.e. users leave the site from the entrance page after viewing only one page
- Exit Rate – the percentage of exits on a page
- Average Session Duration – the average time users (all types) spend on a website
- Goal Completions – show how effective a website is at converting people and getting them to take desired actions
- Goal Conversion Rate – the percentage of users who take a desired action

Google Optimize was used to conduct experiments in order to determine which among the web design variants is best to replace the original web page.

The data-driven web development methodology presented above works only when a live website is available. It is an iterative approach to improving the design of a website page by page while previous methodologies employed the Waterfall approach (Achebe, 2002; Abdul-Aziz et al. 2012). The improvement of the design is primarily based on how the website users respond to the existing web page as reflected in the web analytics analysis. In this way, the needs of the users are given so much importance which is proven beneficial in the work of Achebe (2002) and Flavian et al. (2009).

4.2 Experiments

We show some case studies identified by the web analyst to illustrate how the development methodology works and how the website performs before and after applying data-driven design.

4.2.1 Homepage

The first case study was on the homepage ("Simon Blow Qigong," 2017). It was the first web page that was analyzed as it was obviously the most landed page on the website. The web analyst found out that:

- the homepage does not have a clear call to action for booking a schedule for worships and retreats
- there are a lot of information on the website that cause paradox of choice
- the layout does not have a visual hierarchy in place

From the findings, the hypothesis was formulated: If we clean up the homepage and make the booking of schedule for workshops and retreats more prominent, it is most likely that a lot of bookings will happen. Based on this hypothesis, the web designer created a design prototype and had it approved by the client. With the client's approval, the web developer implemented the design prototype. Figure 1 shows the original homepage on the left and the new homepage on the right.

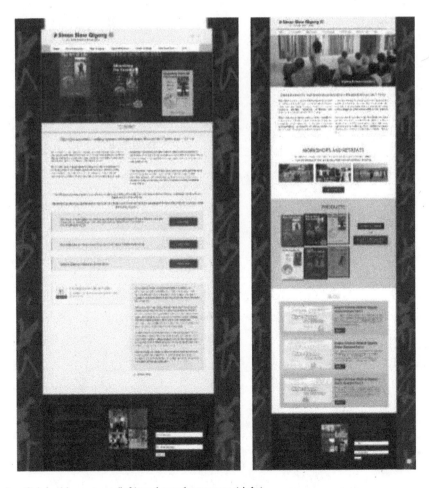

Figure 1. Original homepage (left) and new homepage (right).

The web analyst then ran an A/B test with Google Optimize for 16 days to see how the new page performed over the original. It is shown in Figure 2 that the new homepage is 98% probable to be the best among the design variants and has 98% probability to beat the original page.

Since it is highly probable that the new homepage will yield better result in achieving the web page's goal, it replaced the original homepage on December 4, 2018. Then experiment was done from December 4, 2018 to March 4, 2019 to gather data about the performance of the new homepage in terms of user behavior. The performance of the original homepage was determined during the experiment conducted from September 4, 2018 to December 3, 2018. Although Figure 3 shows that the new homepage has lesser number of users, 0.73% of its users took the desired action of booking a schedule for workshops and retreats, which is an increase of 153.42% from the original homepage's 0.29%. The post-test analysis shows that there is a good sign that the new homepage is working.

4.2.2 Workshops and retreats page

In this case the Workshops and Retreats page ("Simon Blow Qigong," 2017) was redesigned by converting all the links to events on the original page into buttons on the new page (Figure 4).

The new page was implemented on February 18, 2019 and a post test was conducted from February 18, 2019 to April 21, 2019. Post-test analysis shows that the Exit Rate is decreased by 30.65%, the Bounce Rate is decreased by 70.64%, while the Average Session Duration is increased by 19.83% (Figure 5). These results imply that there is an improvement in user click engagements.

Figure 2.   Result of A/B test showing how the new homepage (variant 1) performed over the original.

Figure 3.   Metrics comparison between the original and the new homepage.

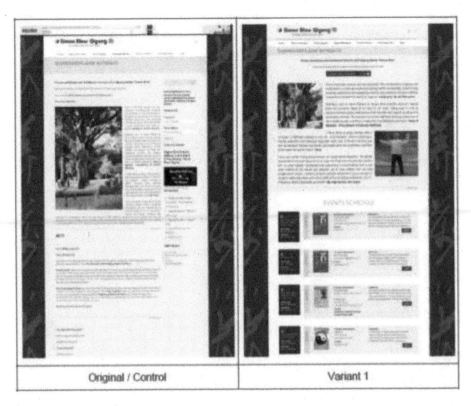

		Original / Control		Variant 1	

Figure 4. Workshops and retreats page.

Landing Page	Sessions	% Exit	Avg. Session Duration	Bounce Rate	Goal Completions	Goal Conversion Rate
	15.06% ↑	30.65% ↑	19.83% ↑	70.64% ↑	100.00% ↑	100.00% ↑
1. /qigong workshops retreats/						
Feb 18, 2019 - Apr 21, 2019	298	26.97%	00:05:54	8.39%	1	0.34%
Dec 17, 2018 - Feb 17, 2019	259	38.85%	00:04:55	28.57%	0	0.00%
% Change	15.06%	-30.65%	19.93%	-70.64%	∞%	∞%

Figure 5. Workshops and retreats page performance.

4.2.3 Qigong Teacher Training page

After analyzing the Qigong Teacher Training page ("Simon Blow Qigong," 2017), the web analyst came up with this hypothesis: If we increase the font size to the same size as the home-page, we will improve the readability of the page and consequently we will get more clicks and an increase in session duration. The hypothesis was tested for four weeks already (March 12, 2019 to April 8, 2019), but the probability to beat baseline for the 'Links Clicked' and 'Session Duration' objectives was still below 95% (Figures 6-7). So, the web analyst calculated the required sample size and duration.

Figure 8 shows that the experiment must be done for 697 days with 1045 samples for it to be able to achieve the confidence level of 95%. Since running the experiment for 697 days is costly and the test is low impact as it is only testing for readability, with the advice of the web analyst, the client decided to have the experiment ended and have the hypothesis implemented

Links Clicked ▾

	Variant ↑	Improvement	Probability to be Best	Probability to beat baseline
☑	Original 50 sessions	Baseline	32%	Baseline
☑	Variant 1 42 sessions	-25% to 59%	68%	68%

Figure 6. A/B test result for the 'Links Clicked' objective.

Session Duration ▾

	Variant ↑	Improvement	Probability to be Best	Probability to beat baseline
☑	Original 50 sessions	Baseline	39%	Baseline
☑	Variant 1 42 sessions	-42% to 110%	61%	61%

Figure 7. A/B test result for the 'Session Duration' objective.

though. This is a case where there is few traffic to the page, thus, running an experiment requires a long time.

5   CONCLUSIONS

In this study, we identified the methodology and standards used in data-driven web development based on an actual website development practice.

The case studies illustrate that data-driven web development is done iteratively. Post-test analysis indicates that web analytics tools have been very useful in generating reports that are insightful in the design process. Further, the experiments reveal that traffic to the website is essential in determining whether a hypothesis brings true improvement. In addition to web analytics analysis, heuristics and usability analyses are necessary in finding some issues on the website. The expertise of the web analysts is very important as he does the heuristics and usability analyses which are experience-based assessments in nature. The heuristics analysis considers relevance, trust, orientation, stimulance, security, convenience, and confirmation while usability analysis uses web usability guidelines.

Figure 8.   Sample size and duration calculation.

As compared to the performance of the original website, the case studies show that data-driven design has improved the performance of the website. However, data-driven approach is time consuming and it can only be done when there already exists a live website.

The case studies being examined used Google Analytics and Google Optimize as the web analytics tools. It would be interesting to explore other web analytics tools (Liikkanen, 2017). Also, qualitative research is another technique that can be done to gather data from the users which may not be revealed by the analytics data. This study evaluated the performance of one website only. Although the present result is favorable, it would be more significant if more websites will be evaluated to obtain a general result. As this study considered only one data-driven web development practice, future work may consider examining other approaches employed by other companies. A research methodology similar to the work of Achebe (2002) and Abdul-Aziz et al. (2012) may be done in order to derive a general data-driven web development methodology.

ACKNOWLEDGEMENT

We gratefully acknowledge the financial support from WebFocus Solutions, Inc. We would like to thank Joven Francis Agno for the insightful discussions.

REFERENCES

Abdul-Aziz, A., Koronios, A., Gao, J. & Suhaizan Sulong, M. 2012. A methodology for development of web-based information systems: web development team perspective. *AMCIS 2012 Proceedings*. 11. http://aisel.aisnet.org/amcis2012/proceedings/SystemsAnalysis/11.

Achebe, I. 2002. *Comparative study of web design methodologies*. (Unpublished master's thesis). Vrije Universiteit Brussel. https://wise.vub.ac.be/sites/default/files/theses/achebe-thesis.pdf

Advanced product analytics shouldn't break the bank. (n.d.). Retrieved from https://www.kissmetricshq.com/.

Data can come from anywhere. (2019). Retrieved from https://www.adobe.com/sea/analytics/adobe-analytics.html.

Domain count statistics for TLDs. (n.d.). Retrieved from http://research.domaintools.com/statistics/tld-counts/.

Flavian, C., Gurrea, R., & Orus, C. 2009. Web design: a key factor for the website success. *Journal of Systems and Information Technology* 11(2): 168–184.

Grow your business by learning why users convert, engage, and retain. (2019). Retrieved from https://mixpanel.com/.

Internet world stats. (2019, July 16). Retrieved from https://www.internetworldstats.com/stats.htm.

Liikkanen, L.A. 2017. The data-driven design era in professional web design. *Interactions* 24(5): 52–57.

Make more money with the data you have. (n.d.). Retrieved from http://www.agnojf.com/.

Nielsen, J. (2001, October 31). 113 Design guidelines for homepage usability. Retrieved from https://www.nngroup.com/articles/113-design-guidelines-homepage-usability/.

Simon Blow Qigong. (2017). Retrieved from https://www.simonblowqigong.com/.

Taylor, J.J., McWilliam, J., Forsyth, H., & Wade, S. 2002. Methodologies and website development: a survey of practice. *Information and Software Technology* 44: 381–391.

The new standard in tracking customer data. (2019). Retrieved from https://heap.io/.

Travis, D. (2016, April 12). 247 web usability guisdelines. Retrieved from https://www.userfocus.co.uk/resources/guidelines.html

Theory and Practice of Computation – Nishizaki et al (eds)
© 2021 Taylor & Francis Group, London, ISBN 978-0-367-41473-3

Analysis on the adequacy of current acceptance criteria in developing scripts for automation testing

Q.J. San Juan
Department of Computer Science, University of the Philippines Diliman, Quezon City, Philippines
Advanced Science and Technology Institute, Department of Science and Technology - Diliman,
Quezon City, Philippines

L.L. Figueroa
Department of Computer Science, University of the Philippines Diliman, Quezon City, Philippines

ABSTRACT: In this paper, the adequacy of the Acceptance Criteria (AC) for software testing was analysed. The set of AC analysed, at the time, is being used in the development of a project in Department of Science and Technology - Advanced Science and Technology Institute (DOST-ASTI). Software testing faces issues in terms of validating test scripts against the AC because in practice, AC is written specifically for use by the developers and not by the software testers in developing scripts for test automation. In order to address this issue, the authors performed a series of data gathering and analysis through questionnaires. These questionnaires evaluate the discrepancy in perspectives of software testers and business analysts to the current AC. It was concluded that the current AC is already adequate for use in writing scripts for test automation, however, there are contents that can be incorporated to the current AC to aid the issue mentioned above. To be specific, there should be indication of the performance level criteria and the necessary data structures in the AC. Performance level criteria will give the software testers information for them to incorporate a threshold in their tests scripts for waiting responses in every action. It was also concluded that data structures must also be explicitly indicated because they give the software testers information on how they will store their test data and how they will manipulate them.

1 INTRODUCTION

Automation testing of software systems is relatively new in software engineering. Even before automation testing, the practice of the industry and the academe has been manual testing. The software testers do the manual testing through creating test cases based from a set of acceptance criteria and then moving forward to writing the test plan. The test plan is a document containing data about the number of software testers available, the overall target duration and costs of testing, the scope, acceptance criteria (Lewis, 2003) (Bowman, 2004). However, since this test plan is based on the acceptance criteria, and it is under the assumption that acceptance criteria is tailored for software developers, this test plan is assumed to be inadequate as guide document by software testers in writing their test scripts. As mentioned earlier, the acceptance criteria document is not entirely written for the purpose of automation testing. This paper will be working on in this issue - the adequacy of the current acceptance criteria being used in a certain project in Department of Science and Technology - Advanced Science and Technology Institute (DOST-ASTI) as guide document for software testers in creating their scripts for test automation.

The authors performed data gathering from the key persons involved in the development of the test scripts for automation which are the software testers, and in the writing of the acceptance criteria which are the business analysts. The data are gathered through the use of questionnaires that assess how the respondents view the current acceptance criteria they are using in addressing the needs demanded by their roles. Later in the paper, their responses will be compared to each other to show how similar and different their perspectives are regarding the adequacy of the acceptance criteria.

The main contribution of this paper are:

1. identify which components in the acceptance criteria need to be improved and added
2. visualize the discrepancy in the software testers' and business analysts' perspective
3. evaluate the relation of user stories' complexity and the overall perception of the respondents

2 REVIEW OF LITERATURE

2.1 *Automation testing frameworks*

As established in the first part of the paper of the history of automation testing, it can be assumed that it is really relative new concept. From the time it was introduced to the present, there have been a number of literature either journal and patents about automation testing already published. Regression testing via manual testing and manual writing of test cases and test scripts are time-consuming and energy-inefficient which makes it more important to use automation testing frameworks to streamline the testing processes. One patented invention produced was an English-based interface for managing test cases that will eradicate the process of coding scripts for test automation and recording the systems being tested (Haswell et al., 2005). Its database is *Microsoft Access 97* or *SQL Server* while its automation tool is *Mercury Interactive WinRunner*.

Another study proposed a new automation testing framework called *TestEra*. *TestEra* is used for automating test of programs written in Java and it automatically creates the test cases and evaluates the program being tested against a set of completeness criteria (Marinov and Khurshid, 2001). For the failed criteria, it produces counterexamples automatically. *TestEra* is written using the Alloy specification language which is a relational-logic based language for writing specifications which makes it fit in the testing of programs following an object-oriented framework (Macedo and Cunha, 2013) such as *TestEra*.

Another patent produced in 2010 was a generic test automation framework that is language-independent and language-agnostic. It is a novel framework that makes the test results readily available to the developers. With it, the developers can access the test results readily via a web service for example (Noller and Mason Jr, 2010). This really provides an agile and innovative way of testing and debugging the system since the developers can check the codes in real time and no delay. Another study proposed an automation testing framework specifically created for JavaScript web applications (Artzi et al., 2011). It is "a framework for feedback-directed automated test generation for JavaScript in which execution is monitored to collect information that directs the test generator towards inputs that yield increased coverage." Another framework was proposed named as *Korat*. *Korat* is a framework for automatic creation of test cases based from a set of predicates and inputs (Boyapati et al., 2002).This paper is specific to Java Programs as the paper implemented the tests in different data structures in Java.

2.2 *Acceptance criteria or specifications*

Specifications of the system can be written in the form of acceptance criteria. Usually, specifications are formal in nature which means it is written using formal language and with formal construction, while acceptance criteria uses rather a more technical way of construction that can be easily understood and transmitted into working codes by the developers. In this paper, one of the methods was the analysis of the adequacy of the acceptance criteria in writing test scripts by the software testers. There have been numerous studies that tackle the software design and specification writing. Most of them established the qualities and usage of a good written specification or acceptance criteria. One mentioned that a formal specification really aids in reducing errors in the implementation of the system and in the same way, it helps guide the software testers in checking the criteria met by the system (Hierons et al., 2009). Another thing that was taken from the paper was that the formal specification can be used as a driving force to further improve the system or not.

Another paper published in 2003 proposed a new way of testing through test input generation based on formal specifications (Offutt et al., 2003). Primarily, the said paper suggested the need for a more formal set of specifications because it describes what the system is ought to do, and in that way, will give software testers the chance to automate its testing more precisely and accurately (Offutt et al., 2003). The said paper presented the specification-based testing which is the opposite of code-based testing. The advantage of a specification-based testing over the code-based testing is that the test cases are already created even before the system is completed. And the test cases were created by just clearly looking at the specifications.

In the above publications, we could already see the relevance of properly writing the specifications and acceptance criteria for it holds a bearing in the overall quality of the system. It does not only hold relevance the development of the system but also in the overall validation and verification of the system through test automation.

What this study can supplement the previously conducted studies is the evaluation of the discrepancy of perception by software testers and business analysts to the acceptance criteria. The authors believed that it is necessary to establish this relation since if we are to make sure that their perceptions to the acceptance criteria, we can minimize the scenarios when software testers create assumptions of the acceptance criteria. This could save up a lot of time of reworking caused by wrong assumptions. In the same way, the generated data could help the business analysts in modifying and strategically writing the current acceptance criteria that could meet the needs of the software testers in test automation.

3 DATA PREPARATION AND COLLECTION

3.1 *Questionnaire*

To collect relevant data, questionnaire method was used. The questionnaire used a 9-point Likert Scale ranging from 1(disagree) to 9 (agree). There was an option NA if the respondents cannot categorize their response between the 9-point scale. All the questions in the questionnaire are positive statements and they are shuffled across the questionnaire. Some of the questions are inspired by (Wilson, 1999) which detailed the parts a formal specification needs to have.

Each questionnaire is composed of 28 questions and these questions can be differentiated into two categories - questions about composition of the AC and questions about AC's content. The composition questions assess how the AC is written - its format, layout, clarity, coherence, understandability, emphasis on relevant parts, sentence construction, appropriate use of words, comprehensiveness of steps and maintainability among others. On the other hand, the content group of questions assesses how relevant the acceptance criteria are, the presence of the key content such as expected output, functional requirements and others. There are 15 questions belonging to the component/composition type and 13 questions falling in the content type of questions.

3.2 *User stories and acceptance criteria*

The system whose AC was analyzed was a system being developed by Department of Science and Technology - Advanced Science and Technology Institute (DOST-ASTI). The scope of this study's analysis was limited to only four libraries in the system namely Allotment Class, Account Group, General Ledger Account, and Sub-Object Code and their major functionalities - Add, Update, Deactivate/Activate, and View List.

The guide document that aided the respondents in answering the questionnaire was prepared to only consist of the four libraries and their corresponding user stories. This document contains the actual acceptance criteria used by the developers and software testers. This is an ever-changing document that as of this writing, the acceptance criteria might already be modified and updated. This document was composed of 16 User Stories (four actions of the four libraries) where each user story has the format:

As <ROLE>, I should be able to <ACTION TO PERFORM>.

Each user story has specific set of acceptance criteria. In Figure 1, the sample criteria show how one can perform the user story, the conditions that need to be met, the restrictions that need to be followed, etc. The length of the acceptance criteria differs in every acceptance criteria. Some have lengthy criteria while some have only short criteria.

3.3 Respondents and context of answering

The team working on the system to analyzed is a small development team - with four developers, three software testers, two business analysts, and one project manager. The current setup of the team is that there is only a single set of acceptance criteria (AC) being shared by both software developers and software testers as guide in writing their development codes and test scripts respectively. Since the AC is written by business analysts, we added them in the respondents of the questionnaire. We included the business analyst in order to compare their viewpoints with that of the software testers in terms of the adequacy of AC.

There were 5 respondents of the questionnaires. Two of which are business analysts and three are software testers. The respondents were chosen since they are the key persons in the development of the system under analysis. Each respondent needs to answer 16 questionnaires corresponding to the number of user stories included in the scope.

The context to be used by the respondents in the answering the questionnaires differed based on their role in the project. For the software testers, they are to answer the questions in the context of translating the criteria into test scripts for automation. For example, if an item reads, "the steps are relevant." It means "are the steps relevant to writing test scripts?".

On the other hand, for the business analysts, they need to answer the items in the context of gathering and verifying the requirements and putting it into acceptance criteria. Following the previous example, the item means, "are the steps included here are relevant acceptance criteria?"

4 ANALYSIS AND RESULTS

4.1 Evaluation of the components of the current acceptance criteria

Insights and valuable data were gathered from the respondents of the questionnaires. The responses were averaged using the scale of the Likert as the weight. In this part of the paper,

---

1. I can create Allotment Class by clicking the New Allotment Class button and by entering the following information: a. Digit * b. Expense Allotment Class * c. UACS Allotment Class Title * d. UACS Allotment Class Name *
Note: Fields with asterisk (*) are required.
2. The following fields shall accept:
  a. Digit – Numbers only (Should be unique. Minimum length: 2 Maximum: 4)
  b. Expense Allotment Class – Alphanumeric Characters
  c. UACS Allotment Class Title * – Alphanumeric Characters
  d. UACS Allotment Class Name * – Alphanumeric Characters
3. A user cannot add Allotment Class without completing all the mandatory fields. If mandatory fields are empty, display a warning message "Fields cannot be blank."
4. When I clicked the Add button, if fields are properly filled in, new entry should be added in the database and I should be able to view created Allotment Class in the Allotment Class List. If fields are not properly filled in, I should not be able to add Allotment Class in the database.
5. Upon successful adding of Allotment Class, I should be able to receive a confirmation message "Successfully added an Allotment Class!"
6. When I clicked the cancel button, the Add Form shall be hidden from the Allotment Class Page.
7. Information from the Add form is stored in the Allotment Class database.
8. Default status is Active.

Figure 1. Sample acceptance criteria (Management information systems unit, DOST-ASTI, 2019).

the discrepancy in the responses of the software testers and business analysts are visualized through the line graphs.

Specifically, the results of conducting the questionnaire answered the following questions:

1. Which components in the current AC score low and high and need to be improved or added?
2. Is there a significant discrepancy between the perspective of business analysts and software testers to the adequacy of the acceptance criteria in their respective contexts?
3. Which among the user stories have low overall scores and high overall scores?

In order to show the difference between the perception of business analysts and software testers about the adequacy of acceptance criteria, the authors provided graphs to show that.

In following Figures 2 and 3,the evaluation of the Composition component and Content component were visualized respectively. The gray area shows the discrepancy between the views of software testers and business analysts in each of the composition component and content component of the acceptance criteria.

In Figure 2, it shows that there is a huge gap in the items *Construction: Appropriate Words* and *Composition: Adjustability*. This could mean that for the business analysts they thought the these components are already adequate and present, when in fact, for the software testers, they thought that these are inadequate for them to write their test scripts.

Figure 3, on the other hand, shows that there is also a huge gap in items *Content:Performance Criteria* and *Content:Data Structures*. There is really a different perception to the acceptance criteria between the software testers and business analysts.

For the figures to follow, we will see the evaluation of only software testers in both the content and composition components of the acceptance criteria showing which of the components have lower scores as compared to the others.

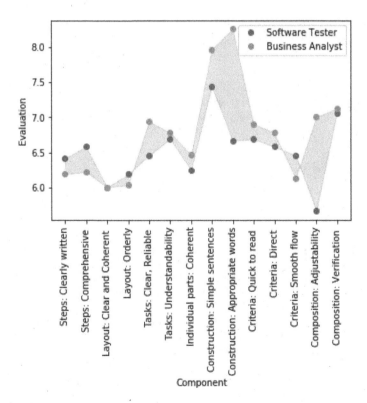

Figure 2.   Evaluation of composition component of the acceptance criteria.

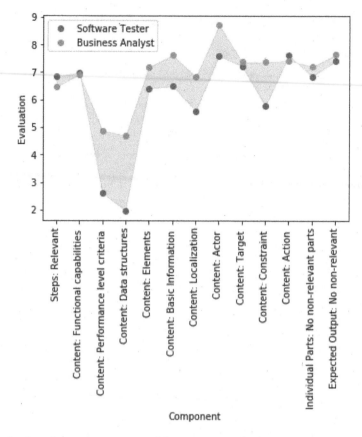

Figure 3.   Evaluation of content component of the acceptance criteria.

In Figure 4, it shows that in terms of composition, there are no extremes in the data which means that the components statistically score the same. There are just some outliers in the data.

In Figure 5, it shows that the *performance level criteria* and *data structures* components score low, while the other components are on the average.

4.2   *Evaluation of user stories*

Apart from the rating based on the components of the acceptance criteria, in this section, it is user stories that are compared to the evaluation of the respondents. In here, it will be shown if there is a trend in the scores among user stories.

The Tables 1 and 2 show the perspectives of software testers and business analysts to the user stories. Specifically, in the perspective of the software testers, the *Create* and *Update* actions are inadequate (low score). With this, maybe, we can assume that it is because there are more conditions that need to be satisfied when testing these actions compared to View List and Activate/Deactivate actions. On the other hand, in the perspective of business analysts, there is no user story having a significantly low score as compared to others. We can conclude that in the perspective of the business analysts, the user stories are equally adequate. However, it is surprising that it is the *View List* actions that score low as compared to the other actions, however, the discrepancies are negligible.

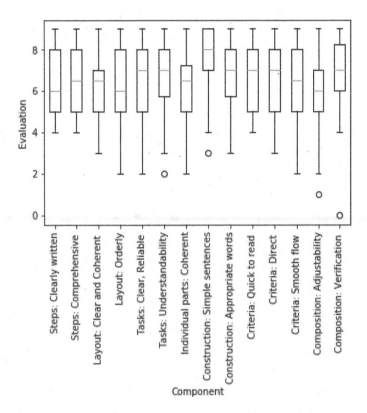

Figure 4. Distribution of responses by software tester to composition questions.

Table 1. Rating per user story by software testers.

User Story	
General Ledger: Create	5.282353
Account Group: Create	5.440476
General Ledger: Update	5.847059
Sub-Object Code: Create	5.870588
Sub-Object Code: Update	6.202381
Account Group: View List	6.411765
Sub-Object Code: Activate/Deactivate	6.445783
Allotment Class: View List	6.576471
Allotment Class: Update	6.600000
General Ledger: View List	6.623529
Allotment Class: Activate/Deactivate	6.674419
Allotment Class: Create	6.825581
Sub-Object Code: View List	6.941176
Account Group: Activate/Deactivate	6.976471
Account Group: Update	7.011628
General Ledger: Activate/Deactivate	7.152941

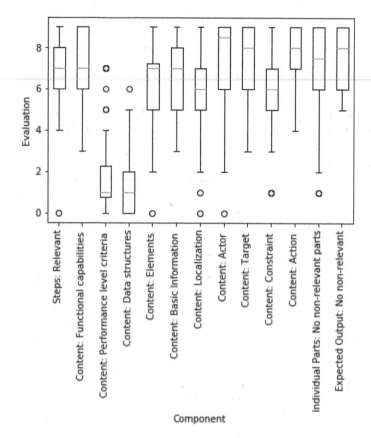

Figure 5. Distribution of responses by software tester to content questions.

Table 2. Rating per user story by business analysts.

User Story	
Account Group: View List	6.103448
Allotment Class: View List	6.310345
General Ledger: View List	6.672414
Account Group: Create	6.810345
Sub-Object Code: Create	6.827586
Sub-Object Code: View List	6.827586
Allotment Class: Update	6.870370
General Ledger: Activate/Deactivate	6.913793
Allotment Class: Activate/Deactivate	6.965517
Account Group: Update	7.017241
Sub-Object Code: Update	7.051724
Account Group: Activate/Deactivate	7.086207
General Ledger: Update	7.103448
General Ledger: Create	7.172414
Sub-Object Code: Activate/Deactivate	7.172414
Allotment Class: Create	7.719298

5 CONCLUSION

It can be concluded that the current acceptance criteria is already adequate for usage by software testers in developing their test scripts. However, there should be some additional content such as *performance level criteria* and as well as *data structures* that need to be incorporated in the current acceptance criteria. Also, some components also need to be worked on because of a satisfactory scores like *localization* and *maintainability*.

Performance level criteria is important since in the test automation, there should be a limit for a test script to run before it produces a result. If there is no threshold for the test scripts in waiting for response, the test scripts will run and run until it already maximizes all of its resources. On the other hand, data structures must also be explicitly indicated in the acceptance criteria because it will give the software testers information to how they can store and manipulate the test data using their scripts. For example, if a certain set of data has a data structure array and it is properly indicated in the acceptance criteria, the software testers can use this information on how they are gonna store their test data and how they are gonna manipulate them.

Gaining insights further from the analysis, it can be concluded that there is a huge gap in the perception of the software testers and business analysts to select components of acceptance criteria. It is in this gap that needs improvement. It could suggest where business analysts can improve more in terms of writing and constructing acceptance criteria that could meet the needs of the software testers in test automation. Another thing to note from the analysis is that was no direct relationship between the type or complexity of user story and the perception of the both business analysts and software testers.

Moreover, there is an implied rule on the usage of the acceptance criteria. There should be nothing left for assumption. All must be clear and easy to understand. For software testing to keep up with the agile process of development, the acceptance criteria should be complete and well-written to avoid repetitive consultations of software testers with business analyst regarding the acceptance criteria, and also to avoid the scenario when software testers create assumptions when writing their test scripts.

This paper does not suggest creation of additional document to contain the acceptance criteria and specifications needed by the software testers. Rather this study proposes to have additional content incorporated in the current acceptance criteria instead.

6 RECOMMENDATION AND FUTURE WORK

A possible extension of this project is to perform the methods in a larger team with more business analysts and software testers. Another future work could be to include the software developers in the respondents as well.

Also, another extended work for this study is to implement the suggested adjustments in the current acceptance criteria and evaluate its actual effect to the overall delivery of the system in terms of time and quality.

ACKNOWLEDGEMENTS

To software testers Wendy, Kate, and Rogel, and business analysts Kristine, Juvy for giving their effort and time to respond to the questionnaires and for giving their generous insights on software testing and acceptance criteria writing, and to DOST-ASTI Knowledge Management Division MIS Unit for giving me the permission to work on this study.I would also like to express my gratitude to DOST-ASTI for the the funding support they have given me.

REFERENCES

Artzi, S., J. Dolby, S. H. Jensen, A. Møller, and F. Tip 2011. A framework for automated testing of java-script web applications. In *Proceedings of the 33rd International Conference on Software Engineering*, pp. 571–580. ACM.

Bowman, J. 2004, April 20. Requirements based software testing method. US Patent 6,725,399.

Boyapati, C., S. Khurshid, and D. Marinov 2002. Korat: Automated testing based on java predicates. In *ACM SIGSOFT Software Engineering Notes*, Volume 27, pp. 123–133. ACM.

Haswell, J. J., R. Young, and K. Schramm 2005, June 14. Language-driven interface for an automated testing framework. US Patent 6,907,546.

Hierons, R., K. Bogdanov, J. Bowen, R. Cleaveland, J. Derrick, J. Dick, M. Gheorghe, M. Harman, K. Kapoor, P. Krause, et al. 2009. Using formal specifications to support testing. *ACM Computing Surveys (CSUR) 41* (2), 9.

Lewis, E. B. 2003, April 8. Technique for automatically generating a software test plan. US Patent 6,546,506.

Macedo, N. and A. Cunha 2013. Implementing qvt-r bidirectional model transformations using alloy. In *International Conference on Fundamental Approaches to Software Engineering*, pp. 297–311. Springer.

Management Information Systems Unit, DOST-ASTI 2019. Sprint document of erp procurement system. Unpublished.

Marinov, D. and S. Khurshid 2001. Testera: A novel framework for automated testing of java programs. In *Proceedings 16th Annual International Conference on Automated Software Engineering (ASE 2001)*, pp. 22–31. IEEE.

Noller, J. and R. Mason Jr 2010, April 6. Automated software testing framework. US Patent 7,694,181.

Offutt, J., S. Liu, A. Abdurazik, and P. Ammann 2003. Generating test data from state-based specifications. *Software testing, verification and reliability 13*(1), 25–53.

Wilson, W. 1999. Writing effective natural language requirements specifications. *Naval Research Laboratory*.

APPENDIX

QUESTIONNAIRE

Items	disagree	1 2 3 4 5 6 7 8 9	agree	NA
1. The steps are written clearly.	disagree		agree	NA
2. The steps are relevant.	disagree		agree	NA
3. The format and layout is clear and coherent.	disagree		agree	NA
4. Functional capabilities are found.	disagree		agree	NA
5. The performance level criteria are found.	disagree		agree	NA
6. The data structures (trees, arrays) needed are specified.	disagree		agree	NA
7. The elements (table, buttons) needed are indicated.	disagree		agree	NA
8. The task is clear and reliable.	disagree		agree	NA
9. The task is easy and quick to understand.	disagree		agree	NA
10. The flow is orderly.	disagree		agree	NA
11. The parts are coherent.	disagree		agree	NA
12. Only relevant parts can be found. If no, please identify the non-relevant parts in a separate paper.	disagree		agree	NA
13. Needed information is emphasized.	disagree		agree	NA
14. Construction of sentences are simple and easy to understand.	disagree		agree	NA
5. Construction of sentences are easy to understand. (Not included in analysis - duplicate of item 14)	disagree		agree	NA
16. The words used are appropriate.	disagree		agree	NA
17. The basic information is already seen.	disagree		agree	NA

(Continued)

Items	disagree	1 2 3 4 5 6 7 8 9	agree	NA
18. The criteria are quick to read.	disagree		agree	NA
19. The criteria are direct.	disagree		agree	NA
20. The flow of criteria is smooth.	disagree		agree	NA
21. The steps are comprehensive.	disagree		agree	NA
22. Items in expected output are relevant.	disagree		agree	NA
23. The criteria have localization.	disagree		agree	NA
24. The actor involved is present.	disagree		agree	NA
25. The target of the step is easily seen.	disagree		agree	NA
26. The constraint of the step is available.	disagree		agree	NA
27. The action of the criteria is available.	disagree		agree	NA
28. Should there be a change in requirement, change is easily added.	disagree		agree	NA
29. Checking if all criteria are met is easy.	disagree		agree	NA

Theory and Practice of Computation – Nishizaki et al (eds)
© 2021 Taylor & Francis Group, London, ISBN 978-0-367-41473-3

Classification of white blood cells using the convolutional neural network architecture

R.J. del Rosario & P.E. Mayol
University of the Philippines Cebu, Cebu City, Philippines

ABSTRACT: Classifying white blood cells is essential in the diagnosis of blood-based diseases. By determining the quantity and relative frequency of blood in the body and comparing it to what is normal, possible blood diseases can be detected and prevented. However, existing processes in diagnosing blood-based diseases pose some problems. They are costly, time-consuming, and inaccurate. Hence, the development of computer-based systems is used to address these problems. In this study, two convolutional neural network models were implemented, namely, the Inception v3 and ResNet model, to classify White Blood Cells (WBC) from around 2,500 microscopic images. Results showed that the Inception v3 model performed better than ResNet model, with an accuracy of 83%, loss of 0.96, average precision of 0.87, recall of 0.84 and f1-score of 0.84 for the former and an accuracy of 74%, loss of 2.09, average precision of 0.81, recall of 0.75 and f1-score of 0.72 for the latter.

1 INTRODUCTION

According to Othman (2017), the quantity of blood cells plays an important role in ensuring the healthiness of a person. By determining it and its frequency in the body and comparing it to what is normal, possible diseases can be inferred. By knowing the possible disease, it can be prevented, and the healthiness of the person can be ensured. Identifying the quantity of the blood cells also requires classifying them. It implies that classifying different kinds of blood cells can also be used as a tool in diagnosis.

Currently, there are several methods on how hematologists diagnose patients with blood-based diseases. These include manual WBC counting by a medical expert, Red Blood Cells (RBC) impedance and using flow cytometry machine counters. The methods mentioned are capable of diagnosing patients with blood-based diseases, however, these methods are costly, time-consuming and prone to error. For these reasons, Maji et al (2015) suggested that the development and use of computer-based systems, instead of traditional methods, will greatly contribute to the acceleration of the analysis process and more accurate results.

There have been several technologies used to improve the accuracy and accelerate the process of classifying WBCs. Image Recognition, which is the ability of machines or software to identify objects, places, people, writing and actions in images, has emerged. Together with deep neural networks, these technologies can classify WBCs through images.

In this study, Inception v3 and ResNet models were used to classify the type of WBC; namely Lymphocytes, Monocyte, Eosinophil and Neutrophil. The results for the models were then compared and evaluated using classification metrics such as accuracy, precision, confusion matrix, f1-score and recall.

2 REVIEW OF RELATED LITERATURE

2.1 *Blood and its components*

The blood has four main components: plasma, RBC, WBC and platelets. Each component has its own role and function. Among the four, the WBC (also called leukocytes) is responsible from protecting the body from infections. This study will focus on WBCs because they are inherently unstable which makes them difficult to classify. There are five types of WBCs: namely, neutrophils, eosinophils, basophils, lymphocytes and monocytes. Table 1 shows the type and structure of

Table 1. WBC types and structures.

Cell Type	Structure
Neutrophil 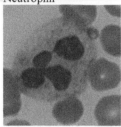	Neutrophil is a granulocyte and contains granules. Its granules are small, numerous and when stained, have a light-pink to bluish-purple, "neutral" color. The cytoplasm of a neutrophil is light pink.
Eosinophil 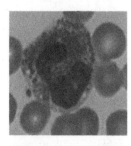	Eosinophil is also a granulocyte and contains granules. Its granules are different from neutrophils. When stained, the large granules becomes red in color. Its granules usually cover the nucleus.
Lymphocyte 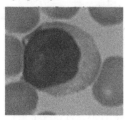	Lymphocytes belong to a mononuclear cell group. Small lymphocytes have a round nucleus and a small amount of blue cytoplasm. The lymphocytes looks very smooth and round, and they also vary a lot. When they encounter an infectious agent, they can appear larger with a lot of cytoplasm. At times, they would appear wavy.
Monocyte 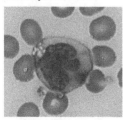	Monocytes also belong to a mononuclear cell group. It usually has a large amount of cytoplasm in relation to the size of the nucleus. They are also usually more irregular in shape than the smooth lymphocytes. The cytoplasm of a monocyte has a dull blue-gray color. It contains granules even though it is not a granulocyte. They appear very fine and lightly stained, giving the monocyte cytoplasm a "ground-glass" appearance.

the WBCs, excluding basophils since it will not be included in the classification due to unavailability of data.

2.2 Related studies

There have been several studies which involve classifying blood cells from microscopic images using neural network. In classifying blood cells from microscopic images, Ongun et al. (2001) applied the multilayer perceptron network trained using conjugate gradient descent (CGD), linear vector quantisation (LVQ) and K-Nearest Neighbor (KNN) classifier which produced 89.74%, 83.33% and 80.76% of accuracy, respectively.

Ramoser et. al. (2005) developed a segmentation method for an automated blood differential counting using image analysis. The segmentation method is used in the study to extract WBC and separate its constituents which are nucleus and cytoplasm. The study achieved an overall average error of less than 8%.

Another study by Dorini et al (2012) also used segmentation method for automated differential counting. The study performed a two-level segmentation method; first is to segment the nucleus using SMMT filter, and second, to identify the cytoplasm region using granulometry and morphological operations. WBC types are then classified using a KNN classifier with geometrical shape features and obtained a reasonable accuracy (78% performance vs 85% classified manually by a specialist). All the studies previously mentioned deal with feature extraction, WBC segmentation and classification.

In the study by Habibdazeh et. al. (2013), they performed automatic classification of WBC cells into one of five WBC types in low resolution cytological images using 3 classifiers. The data used in the study were low resolution blood smear images containing various blood cells in addition to WBC. From the images, WBCs were separated using image processing. As mentioned, three classifiers were used to perform automatic classification of WBC cells. Two of the classifiers were SVM-based with different configuration. One of the classifiers used was Convolutional Neural Network (CNN). CNN was different among the two classifiers and the neural networks used in the previous studies. Unlike the SVM-based classifiers, CNN does not need prior knowledge on the data. Despite that, it was able to classify WBC best among the two SVM-based classifiers with recognition rates either higher or comparable to the SVM-based classifiers for all five types of WBCs.

CNN-based classifiers often cannot process raw images directly. However, it has automatic methods which can retrieve features directly from raw data which is generally preferable. Because of this feature, CNN-based classifier can solve classification problems without prior knowledge on the data.

3 METHODOLOGY

3.1 Data gathering

The dataset used is located in Paul Mooney's repository in Kaggle (https://www.kaggle.com/paultimothymooney/blood-cells), of which the original raw data is hosted on Shenggan's BCCS Dataset Github page (https://github.com/Shenggan/BCCD_Dataset) and is under MIT license. The repository contains Blood Cell Images of 4 different cell types; namely eosinophil, lymphocyte, monocyte, and neutrophil. The dataset contains two main directories (TRAINING and TESTING) and under these two main directories are directories named after WBC types. The subdirectories under TRAINING contain about 2500 augmented images and subdirectories under TESTING contain about 620 augmented images.

3.2 Data preprocessing

The images dataset was loaded into Jupyter notebook. For compatibility and speed purposes, all the images were resized into 75x100 before stored into NumPy array. Along with the

Table 2. WBC labels and class.

Label	Class
1	Eosinophil
2	Lymphocyte
3	Monocyte
4	Neutrophil

images, the label for each of the data is stored as well, in a NumPy array. After that, four NumPy arrays were generated; one for training images, one for training labels, one for testing images, and the last one for testing labels.

3.3 Design

Model specifics were finalized such as the input features, input format, packages used for building the models, number of training epochs, and classification metrics which determined model performance.

The Inception v3 model and ResNet models were designed to accept images and labels as inputs. The training images were of the shape (9957, 75, 100) indicating the total number (9957) and the width (75) and the height (100) for both models. The testing images were of the shape (2487, 75, 100) for both models as well. The training and testing labels are an array of integers and both have lengths of 9987 and 2487 respectively. Each label represents WBC type and corresponds to each of the images in the training and testing data. Table 2 shows the label and the class of WBC the image represents.

The models used which are Inception v3 and ResNet models were available at the Keras library. Since these were used as classifiers and feature extractor, all layers were set to trainable. The models were then configured for training.

3.4 Code generation and model training

In this phase, the source code for the model implementation was designed and created in Python using Keras library for Inception_v3. The model was then trained on a training set spanning 80 percent of the dataset for each class to classify the type of white blood cells.

3.5 Testing and evaluation

During the testing phase, the model was evaluated against the test dataset, the remaining 20 percent of the dataset for each class. The accuracy of the model was then computed.

4 RESULTS AND DISCUSSION

4.1 ResNet model

Using the same machine to train the images for both, ResNet took 7.47 hours to train around 12,500 microscopic images of WBC. It averaged around 841 seconds per epoch.

Figure 1 shows graphs of the learning curves of the ResNet model in classifying WBC types. The results show that the model is performing well in terms of accuracy, albeit a bit overfitted. The training and test accuracy increase as the number of epoch increases. The test accuracy is also less precise compared to training accuracy. The model was evaluated and yielded an accuracy of around 75%.

Figure 2 shows the actual value of correct and incorrect classification of WBC types using ResNet model. It shows that Lymphocyte yielded the highest number of correct

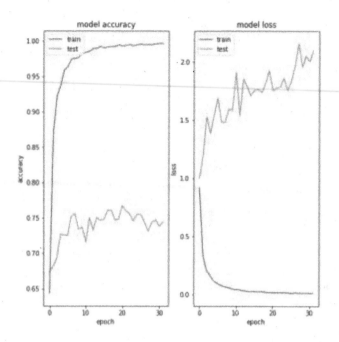

Figure 1.   ResNet learning curve.

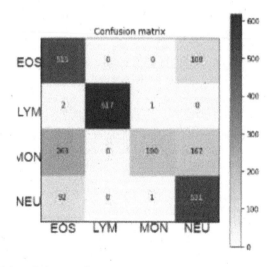

Figure 2.   ResNet model confusion matrix.

classifications, with a value of 617, followed by Neutrophil, Eosinophil and Monocyte, with values 531, 515, and 190, respectively. In terms of precision, lymphocyte attained the highest value, with 617 correctly classified out of 617 total classification, while Eosinophil attained the lowest, with 515 correctly classified out of 872 total classification.

Table 3 shows the classification metric derived from Figure 1. The results show that Lymphocyte is the WBC type that is classified best by the model having the highest number of correct classification among all other WBCs. It attained a precision level of 1.0. Eosinophil, on the other hand, had the lowest precision at precision level of 0.59.

Table 3. ResNet classification metric.

	precision	Recall	f1-score	recall
EOSINOPHIL	0.59	0.83	0.69	623
LYMPHOCYTE	1.0	1.0	1.0	620
MONOCYTE	0.99	0.31	0.47	620
NEUTROPHIL	0.66	0.85	0.74	624
Weighted Average	0.81	0.75	0.72	2487

4.2 Inception v3 model

The model took around 5.12 hours to train around 12,500 microscopic images of WBC. It averaged around 576.5 seconds per epoch. It took faster training time compared with ResNet.

Figure 3 shows a graph of the learning curve of the Inception v3 model in classifying WBC types. In terms of accuracy, the results show that the model is performing well. The training accuracy increases as the number of epoch increases. The same goes for the test accuracy. However, the accuracy values from test set are not as precise as the values from the training set. The results also show a large gap between the test loss and training loss. Both of the curves show improvement; however, the test loss is not as precise as the training loss. The model was evaluated and yielded an accuracy of around 83%.

Figure 4 shows the actual value of correct and incorrect classification of WBC types. Lymphocyte yielded the highest number of correct classifications, with a value of 609. It is followed by Neutrophil, Monocyte and Eosinophil, with values 577, 465, and 430, respectively. In terms of precision, Lymphocyte attained the highest value while Neutrophil attained the lowest. The model yielded a correct classification value of 609 out of 609 total classifications for Lymphocytes. For Neutrophil, the model only classified 577 out of 919 total classification correctly. Lymphocytes yielded the most precise classification as it has the most distinct features among the other types. The other three, on the other hand, shared common characteristics and features thus achieving less precision.

Table 4 shows the classification metric derived from Figure 3. The results, using the Inception v3 model, show that Lymphocyte is the WBC type that had the highest count of correct classification. It attains a precision level of 1.0. Neutrophil, on the other hand, had the poorest result, having a precision level of 0.63.

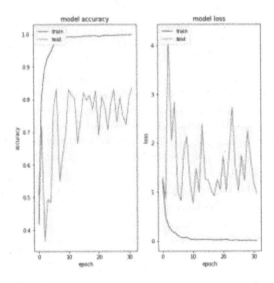

Figure 3. Inception v3 learning curve.

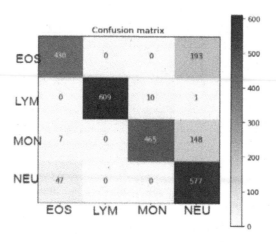

Figure 4. Inception v3 confusion matrix.

Table 4. Inception v3 classification metric.

	precision	recall	f1-score	recall
EOSINOPHIL	0.89	0.69	0.78	623
LYMPHOCYTE	1.0	0.98	0.99	620
MONOCYTE	0.98	0.75	0.85	620
NEUTROPHIL	0.63	0.92	0.75	624
Weighted Average	0.87	0.84	0.84	2487

4.3 *ResNet vs inception*

In terms of training runtime, Inception v3 model performed faster with total training time of around 5.12 hours while the ResNet model took around 7.47 hours on 12,500 images. Inception v3 took faster training runtime due to its more optimized layer.

In terms of model accuracy and model loss, Inception v3 model performed better compared to the ResNet model. The Inception v3 model had an accuracy of about 83% and model loss of about 0.96, while the ResNet model, about 74% and 2.09, respectively. The Inception v3 model yielded a better accuracy due to its architecture. Inception v3 has a more optimized convolution layer and utilized more auxiliary classifiers. On the other hand, ResNet model only made use of residual block to boost the performance of deep neural network.

In classifying each type of WBC, both models yielded the highest precision in classifying Lymphocyte. Both models attain a precision level of 1.0. On the other hand, Eosinophil had the poorest results for ResNet and Neutrophil for Inception v3. As explicit feature extraction was not performed, it is safe to assume that these results are due to the image datasets for all WBC types. Among the WBC types, Lymphocyte yielded the highest precision because it has the most distinct feature. Its physical appearance looks smooth and round. On the other hand, the two types, Eosinophil and Neutrophil, which yield the poorest results share the same characteristics. Both are granulocytes and characterized to have granules. The main difference for both is the color of the granules. In terms of shape and structure, both types appear to be similar thus confusing the CNN models.

Based on classification metrics, Inception v3 model also performed better in terms of average precision, recall and f1-score of 0.87, 0.84 and 0.84 respectively. ResNet model, on the other hand, yielded values of 0.81, 0.75, and 0.72 respectively. This implies that given similar datasets, the Inception v3 model outperformed the ResNet model in different classification metric areas.

5 CONCLUSION

The classification of type of White Blood Cells is possible given the right input features. The models used in the study, Inception v3 and ResNet, were able to classify White Blood Cells Types from microscopic images.

In this study, the results show that the Inception v3 model performed better than the ResNet model give the input data. As shown in the result, it was found that Inception had a faster runtime compared to ResNet. This might not always be the case given a different dataset with more defined features. Inception v3 had an accuracy of about 83% and a model loss of about 0.96, while the ResNet had an accuracy of about 74% and a model loss of 2.09. In terms of precision, recall and f1-score, Inception v3 also outperformed the ResNet.

An automated classifier can be used by hematologists and other medical professionals dealing with microscopic WBC images to aid them in their manual laborious task of classifying WBC.

6 RECOMMENDATIONS

For future researchers, it is recommended to further optimize the model to increase its accuracy. It is recommended to explore these ways in classifying WBC from images. It is also recommended to gather more data in order to fully maximize the potential of the model. In the study, the WBC types Eosinophil and Neutrophil shared the same characteristics which yielded the poorest result in terms of precision. To further improve the accuracy of the classification of the different WBC, it is recommended to consider possible features of different types of WBCs that could be determined and used to train. When possible features are predetermined, accuracy on closely related datasets can also be improved. Lastly, it is recommended to use the model to classify WBC in real time, via capturing each type of WBCs behavior via video, as this can be used as an additional feature to classify the different WBCs. In the future, it may be possible to gather videos of WBCs in action. Results should also be compared to the conventional methods used by hematologists to determine its viability.

REFERENCES

American Society of Hematology. 2019. Blood Basics. Retrieved September 21, 2019, from https://hema tology.org/patients/basics/.

Dorini, L.B., Minetto, R., Leite, N. 2012. Semi-automatic white blood cell segmentation based on multi-scale analysis. IEEE Transactions on Information Technology in Biomedicine.

Elen, A and Turan, M. 2019. Classifying White Blood Cells using Machine Learning Algorithms. Uluslararasi Muhendislik Arastirma ve Gelistirme Dergisi. 141-152. doi: 10.29137/umagd.498372.

Habibzadeh, M., Krzy'zak, A. and Fevens, T. 2013. White blood cell differential counts using convolutional neural networks for low resolution images. In Artificial Intelligence and Soft Computing, pages 263–274. Springer Berlin Heidelberg.

Maji, P., Mandal, A., Ganguly, M. and Saha, S. 2015. An Automated Method for Counting and Characterizing Red Blood Cells Using Mathematical Morphology. IEEE International Conference on Advances in Pattern Recognition, Kolkata, pp. 1–6.

Ongun, G., Halici, U., Leblebicioglu, K., Atalay, V., Beksac M., and Beksac S. 2001. Feature Extraction and Classification of Blood Cells for an Automated Differential Blood Count System. In Proceedings of the International Joint Conference on Neural Networks, USA, pp. 2461–2466.

Othman, M., Mohammed, T., Ali, A. 2017. Neural Network Classification of White Blood Cell using Microscopic Images. International Journal of Advanced Computer Science and Applications, 8(5). doi:10.14569/ijacsa.2017.080513. Retrieved December 1, 2018, from https://pdfs.semanticscholar.org/bf3b/8d397f4d3306c03ed1df9065c1aeeda2ffc4.pdf.

Ramoser, H., Laurain, V., Bischof, H., Ecker, R. 2005. Leukocyte segmentation and classification in blood-smear images. In Proceedings of the 27th IEEE Annual Conference Engineering in Medicine and Biology, Shanghai, China, September 1-4, pp. 3371–3374.

Rouse, M. 2017. What is image recognition? - Definition from WhatIs.com. Retrieved December 1, 2018, from https://searchenterpriseai.techtarget.com/definition/image-recognition.

University of Wisconsin (2019). White Blood Cells – Medical Technology. Retrieved September 21, 2019, from https://www.uwosh.edu/med_tech/what-is-elementary-hematology/white-blood-cells.

Theory and Practice of Computation – Nishizaki et al (eds)
© 2021 Taylor & Francis Group, London, ISBN 978-0-367-41473-3

Software engineering in the Philippines: A survey and analysis of the academe and industry practices

G. Tongco-Rosario, C. Estabillo & L.L. Figueroa
University of the Philippines, Diliman, Quezon City, Philippines

ABSTRACT: This paper studied the state of software engineering practices in the Philippines. Survey, analysis of projects and interviews were conducted in the industry as well as the academe. Based on the survey results, most companies use agile method with scrum as the most preferred variant. Waterfall, prototyping and timeboxing are also evident in these companies, at times used together (e.g agile scrum-waterfall). The most common method used by students in software development projects is waterfall, followed by prototyping. There is a general awareness of agile-scrum in the universities studied, with the method implemented in one of the student projects. For companies, time-constraint and re-work due to requirement change or additional requirement remains to be the most common challenge. Planning in using agile method and frequency of scrum meetings can also be improved. In the academe, risk management can be complemented with issue management as a way to handle realized risks.

1 INTRODUCTION

IEEE Standard Glossary of Software Engineering Terminology defined Software Engineering (SE) as: (1) The application of a systematic, disciplined, quantifiable approach to the development, operation, and maintenance of software; that is, the application of engineering to software. (2) The study of approaches as in (1) .

Current software engineering model processes include scrum and extreme-programming, both agile processes. Distinguishing between these two is challenging since they share common principles like iteration and production/release of working software. In practice, these two differ in a lot of areas like length of iteration, client-involvement in the project, and incorporation of basic engineering practices (e.g. pair programming and use of automated testing). Other software engineering process models include waterfall, prototyping, spiral model and time-boxing. According to (Raval and Rathod, 2013), Waterfall Method is the oldest and widely-used SDLC (Software Development Life Cycle) model. It focuses on identifying the goal and requirements at the start of production. The spiral model is an incremental model that focuses on risk analysis. Jalote et. al. described timeboxing model as an iterative process wherein each time box, its unit of development, contains all requirements/features needed for the release of one iteration. Prototyping, on the other hand, is another incremental model but its focus is on building and exercising the prototype parts to determine the best implementation of the desired product (Boehm, Gray, &Seewaldt, 1984).

In 2006, Sison et. al. has conducted an exploratory study on software practices in five ASEAN countries namely Malaysia, Philippines, Singapore, Thailand and Vietnam. The authors conducted a sampled software development firm- defined as a for-profit organization whose main business is the development of software for customers external to the organization. Included in this definition are developers of generic software products (e.g., Microsoft) as well as providers of software solutions for specific domains (e.g., Infosys). Excluded are IT departments that cater primarily to the needs of the organizations of which they are part. The authors used Software Engineering Institute Software Capability Maturity Model (SEI SW

CMM) Maturity Questionnaire for Level 2. In this study, specific software practices were assessed. Another study conducted by Sison et.al. is the use of agile methods and practices in the Philippines. Here the authors identified two software firms that were using the agile methodologies as well as the level of usage and benefits derived.

Study of software engineering practices in the US and Japan have been conducted as early as 1984 (Zelkowitz, Yeh, Hamlet, Gannon & Basili, 1984). Recent studies have been done in Turkey by Garousi et.al. (Garousi etal, 2015 & Garousi etal, 2016). There have also been software engineering studies focusing on the academe-industry collaborations (Garousi, Petersen & Ozkan, 2016 & Sandberg &Crnkovic, 2017). Use of agile methodologies in this context (academe-industry) have been probed.

In the Philippines, software engineering is taught in the academe as part of the Computer Science curriculum. In the industry, organizations usually prescribe and/or train the employees on the software engineering practices to use in developing software.

According to an executive report, IT-BPM Roadmap 2022: Accelerate PH, Future Ready on December 2017, Software Development is one of the most established sector of the Philippine IT industry making up the greatest chunk of 28% of the IT sector revenue. IT sourcing industry registered US$2.9 Billion in 2016 and is expected to grow to US$5.7 Billion by year 2022.

It is beneficial for the academe and industry to have an understanding of the software engineering practices taught and used to ensure optimum collaboration. For this purpose, the researchers conducted a study of the state of software engineering practices in the country. The goals of the study are to answer the following questions:

- What is the current software development process used in the academe and industry?
- How strict does the organization follow the Software Development (SD) process?
- What are the common challenges with the adopted SD process?

We present the methodology used, results and discussions, and conclusions and recommendations in the succeeding sections.

2 METHODOLOGY

Three methodologies were used to gather data for the study: survey questionnaires, interviews and analysis of implementation of student projects in SE classes. Survey questionnaires in Google form are sent via social networking site to people working in IT software industry. Due to limited time, purposive sampling is used to determine the respondents for the said survey questionnaires. More than 100 people were invited to answer the google form, however, only 89 answered completely. The distribution of respondents can be seen in Table 1.

The questionnaire is composed of three sections. The first section asked for some basic information about the respondents' company, job and process model that they are using in current projects. To clarify on the software engineering process models, diagrams for each model are attached to it. The choices are Waterfall, Agile – SCRUM, Agile – XP (Extreme Programming), Spiral, Timeboxing and Prototyping. The second section is 4-point Likert-type questions consisting of ALWAYS, OFTENTIMES, SOMETIMES, NEVER. It consists of 17 questions based on the characteristics of the different SE process models given. The third section asked about the challenges they faced when using the current model in Software

Table 1. Distribution of respondents.

Category	Number
Traditional Companies	71
Start-up Companies	8
Students	10

development and their opinion on what is the best process model to be used. Since the question is qualitative, the researchers classified the given challenges so that those with the same idea are grouped together.

To assess the strength of reliability of the items in the questionnaire, Cronbach's alpha was computed using the formula.

$$\alpha = \left(\frac{k}{k-1}\right)\left(1 - \frac{\sum_{i=1}^{k} \sigma_{y_i}^2}{\sigma_x^2}\right) \tag{1}$$

where k refers to the number of scale items; $\sigma_{y_i}^2$ refers to the variance associated with item i; and σ_x^2 refers to the variance associated with the observed total scores (De Vellis, 1991). According to (Christensen, Johnson &Turner, 2011), a high value of the Cronbach's alpha would mean that the items are consistently measuring the same thing, thereby showing an internal consistency reliability.

The researchers have also interviewed start-up companies to further probe the software engineering practices. The interviewees were respondents to the survey and as such, questions during the interview focused more on deep dive and clarifications. Interviewees were asked to describe their usual process on how they proceed with their software product development starting from conception to deployment. The researchers clarified on the challenges and setbacks they encountered using the process they use and the reason for their selection of this model.

For the analysis of the software engineering practices in the academe we looked at two universities offering software engineering courses. We studied the course offering and requirements and looked at the actual execution based on the documents provided. For some of these projects, members were also requested to participate in the survey. Effort to include further samples (e.g. more universities) have been done. However, it was not successful and is thus reserved for further study.

Insights on two more companies (A, B) was also provided based on actual work experience complemented with surveys.

3 RESULTS AND DISCUSSION

3.1 Survey results

Cronbach's alpha analysis to determine the strength of the reliability of the survey items resulted in an alpha value of 0.67.

Given that there are a few number of items for the survey, the researchers deemed that a value of 0.67 is acceptable for this study.

It can be observed in Figure 1 that there are two highly used models in the traditional/corporate setting. These are agile-scrum and waterfall model. The highest number of respondents answered Scrum, followed by combination of Scrum and Waterfall, then purely Waterfall model.

There are 58 respondents who answered agile-scrum whether purely scrum or with other process model which amounts to almost 82% of the total respondents.

Only one answered time boxing as their company's process model showing that time boxing model as described in (Alshamrani and Bahattab, 2015) is very rarely used in the Philippines. The same applies to spiral model. The challenge that the respondent indicated when using this process model is the difficulty in finding an alternative solution for a possible risk. However, when asked what process model he prefers given this challenge, he still responded spiral model.

Figure 2 shows that all respondents who are currently in a start-up company answered agile-scrum as their main process model in developing software. 25% added prototyping as an additional model in their software development process.

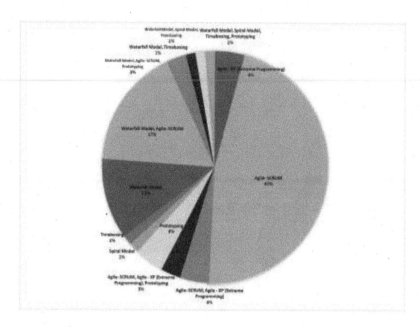

Figure 1. SE models used in corporate/traditional company setting.

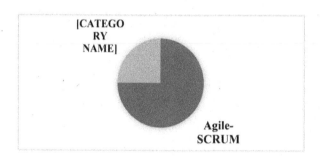

Figure 2. SE models used in start-up companies.

Being in a start-up company, most of them believe that the fast-paced characteristic of this scrum process makes it a good model for the fast-growing industry of software development.

The academe setting, however, shows a different scenario. As we can observe in Figure 3, most student projects are being done using waterfall model while some uses prototyping. It should also be noted that students who responded waterfall belongs to University B while those who answered prototyping is from University A.

Table 2 shows the percentage of response given by the respondents in the Traditional/Corporate setting. Q2 (Integration of testing throughout the lifecycle) got the highest response as an SE practice that is ALWAYS done in a traditional/corporate setting. This is followed by Q17(Following design/coding standards set by the management), Q7(Goal is set at the start of development), Q12(Team identifies the features of project to prioritize and deliver first) and Q8(Project development is iterative). All 5 practices belong to Agile process except for Q7 which is more on the Waterfall model process. The least practice done among the given choices were Q14 (Paired-programming) which is an Agile-XP practice followed by Q4 (Requirements are defined and wouldn't be changed) and Q10(Developing a prototype at the start of project) which is a prototyping SE practice.

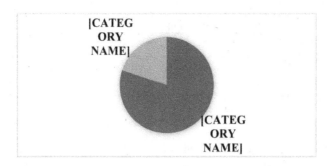

Figure 3. SE models used in academe setting.

Table 2. Survey question response in traditional companies.

	QUESTION	A	O	S	N
Q1	Our project client is very involved in the product development process. (Client is present in every meeting)	30%	37%	30%	4%
Q2	We integrate testing throughout the lifecycle, enabling regular inspection of the working product as it develops.	73%	21%	4%	1%
Q3	New features are delivered quickly and frequently to our client.	46%	42%	11%	0%
Q4	Our requirements are defined and these wouldn't be changed.	11%	41%	38%	10%
Q5	We have highly disciplined process of documentation.	44%	41%	13%	3%
Q6	We follow a strict linear sequence: requirements, design, implementation, testing, and deployment.(no iteration)	32%	30%	24%	14%
Q7	The goal is set at the start of development.	65%	28%	4%	3%
Q8	Our project development is iterative	62%	27%	11%	0%
Q9	Each iteration is done in a timebox of fixed duration.	51%	34%	15%	0%
Q10	We develop a prototype of the Software product at the start for the client's approval.	31%	23%	31%	15%
Q11	We start a project by doing risk analysis (identify and resolve risks)	35%	39%	20%	6%
Q12	Our team identify features of project to prioritize and deliver first.	65%	24%	8%	3%
Q13	The client identifies features of project to prioritize and deliver first.	44%	39%	14%	3%
Q14	We practice paired-programming.	20%	28%	34%	18%
Q15	We have tools for automated testing.	48%	23%	17%	13%
Q16	Our team typically work no longer than 1-2 weeks long in one iteration.	30%	35%	25%	10%
Q17	We follow design/coding standards set by the management.	68%	20%	11%	1%

Note: A=Always, O=Oftentimes, S=Sometimes, N=Never

Table 3 shows the percentage of response given by the respondents from the Start-up companies. Q2(Integration of testing throughout the lifecycle), Q8(Project development is iterative), Q9(Each iteration is done in a timebox of fixed duration) and Q16(Sprint that is no longer than 1-2 weeks) got the highest ALWAYS response. All such SE practices belong to Agile process, although, Q16 is a characteristic leaning towards XP instead of Scrum.

It is quite noticeable, although expected, that the least practiced among the choices in the start-up companies is Q6(Following a strict linear sequence of SE product development without iteration) since this is the number 1 characteristic of Waterfall model.

As shown in Figure 4, the highest number of response we got for the question on the challenges encountered using their particular process model is related to the very fast-paced demand on requirements. Many believed that continuous changing of the client's requirement hinders their productivity since most of the time they have to rework what has already been done. It is quite noticeable that this sentiment is shared by those who are using Agile-scrum and Waterfall model.

Table 3. Survey question response in start-up companies.

	QUESTION	A	O	S	N
Q1	Our project client is very involved in the product development process. (Client is present in every meeting)	38%	25%	38%	0%
Q2	We integrate testing throughout the lifecycle, enabling regular inspection of the working product as it develops.	63%	38%	0%	0%
Q3	New features are delivered quickly and frequently to our client.	50%	38%	13%	0%
Q4	Our requirements are defined and these wouldn't be changed.	13%	38%	50%	0%
Q5	We have highly disciplined process of documentation.	25%	63%	13%	0%
Q6	We follow a strict linear sequence: requirements, design, implementation, testing, and deployment.(no iteration)	0%	13%	50%	38%
Q7	The goal is set at the start of development.	50%	25%	25%	0%
Q8	Our project development is iterative	63%	25%	13%	0%
Q9	Each iteration is done in a timebox of fixed duration.	63%	25%	0%	13%
Q10	We develop a prototype of the Software product at the start for the client's approval.	38%	13%	38%	13%
Q11	We start a project by doing risk analysis (identify and resolve risks)	25%	25%	38%	13%
Q12	Our team identify features of project to prioritize and deliver first.	38%	63%	0%	0%
Q13	The client identifies features of project to prioritize and deliver first.	13%	25%	50%	13%
Q14	We practice paired-programming.	25%	38%	25%	13%
Q15	We have tools for automated testing.	38%	25%	25%	13%
Q16	Our team typically work no longer than 1-2 weeks long in one iteration.	63%	38%	0%	0%
Q17	We follow design/coding standards set by the management.	38%	50%	0%	13%

Note: A=Always, O=Oftentimes, S=Sometimes, N=Never

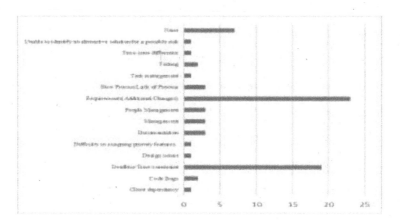

Figure 4. Challenges in SLDC in traditional/corporate setting.

Another challenge is the tight deadlines and short time frames. Some functionalities are difficult to accomplish with the given time frame. For some who are using Agile-scrum, such fallback are usually treated as a technical debt and passed on to be done on the next sprint although some would still have to sacrifice their time and conduct an overtime.

A word cloud showcasing the qualitative answers of the respondents on the challenges they usually encounter is shown in Figure 5.

Both problems arising from these challenges on requirement-change and time-constraint should have been addressed by the agile methodology. Having daily scrum meetings and sprints should have eliminated the case of overtime and increase the productivity. However, this is not the case.

Figure 5.   Word cloud of raw data on challenges.

The same result on problem of overtime was obtained in (Sison & Yang, 2007). Reductions in overtime was not yet achieved despite using daily scrum meetings because of underestimated amount of time for the sprint. In fact, some developers claim that daily scrum meetings are taking too much valuable time when they could be working on their task already.

3.2   Software engineering in the academe

3.2.1   Software engineering course offering

In the Philippines university students are exposed to software engineering as part of the Computer Science course offering. In the undergraduate degree these are divided into two subjects (SE 1 and SE 2) taken for two semesters (e.g. University of the Philippines-Diliman, UP Manila, University of Santo Tomas to name a few). This is also true in both of the university samples looked at. Some universities provide the course in two parts: theory and laboratory (e.g. MAPUA). Advanced software engineering topics are also offered in the graduate degree.

For UP Diliman (University A), software engineering 1 course covers the principles of software engineering, software project management, requirements engineering, software architecture and design patterns, software quality assurance, and software testing. The subject requirements include project documentation deliverables, requirements engineering deliverables, design engineering deliverables, and testing deliverables. Software engineering 2 focuses on software implementation and maintenance. The course requirement includes implementation and testing deliverables, alpha/beta analysis and testing deliverables, at least two (2) software project releases, final project and oral defense.

For University B, the software engineering course examines how systems are developed and the process it takes to propose one to an actual client. A project proposal is required, intended for later development and implementation in SE 2. SE 1 requirements included: (a) System Project Management Plan, (b) System Requirement Specifications, (c) System Design Document and (d) System Test Plan. SE 2 carry on requirements of SE 1. It emphasizes the technical foundations of software development, such as requirements analysis, cost estimation, design, team organization, quality control, configuration management, verification, testing, software implementation, and maintenance. SE 2 course requirements focused on the project deliverables (i.e. project documents and project defense).

For University A, the software process model is discussed in software engineering 1. These process models include the waterfall, the V model, incremental model, evolutionary models (i.e. prototyping, spiral, concurrent), and other process models (e.g. component based development, formal methods, unified process).

3.2.2 Nature of the project and project team

Software engineering courses requires students to practice and implement what they have learned. As such, part of the requirement is to have a software development project.

In this study, we looked at 9 undergraduate student projects in the universities. Seven of the projects came from University A while two came from University B. Each of these project teams is composed of three to five members. University A students maintained a project blog where they upload the required project documents for the course, weekly (most times, with some additional requirements on some weeks). For University B, the researchers were provided final documentations.

3.2.3 Software engineering practices

We looked at the software engineering practices used in the student projects. Due to the nature of the course and its requirements, the default model to use was waterfall. Survey results show that University A students respondents are using prototyping while University B are using waterfall. Details on what we have found in terms of the software engineering practices used are provided below.

3.2.3.1 UNIVERSITY A - WATERFALL AND PROTOTYPING

At the beginning of the SE 1 course for University A, the students were required to identify a problem to solve, applying the software development process. Then each week, specific deliverables are submitted. In general, the documents and deliverables had a standard template. We confirm that the students are using the waterfall model evident in documentations produced and the sequence of activities performed. A prototype was also created across the seven projects. We provide more insights into the specific documents and activities performed by students in University A below.

A. Project Documentation - The project documentation covers the customer requirements, plan of work, schedules and milestones, project metrics, effort breakdown and project risks. It can be observed that some project teams have not provided the mitigation in the initial version of the document.

B. Requirements Visualization - The requirements were visualized using different diagrams. A use case diagram, system sequence diagram, class associations and dependencies diagram, swim lane diagram and entity relationship diagram were utilized.

C. Architectural decisions - Each system architecture decision were stated as follows: Subject, Architectural Decision, Issue or Problem Statement, Assumptions, Motivation, Alternatives, Decision, Justification, Implications, Derived Requirements, Related Decisions. Status of these decisions were also summarized.

D. Prototype - A prototype was produced 4 weeks after the customer requirements were derived and modeled. The prototype was built to produce a more detailed technical requirement and design. Here the students seem to adopt the prototyping approach as an alternative to specifying (Sison, Jarzabek, Hock, Rivepiboon, & Hai, 2006). This is consistent with the University A students' survey response. After the prototype was created, detailed requirement and design were produced.

E. Requirements Model and Architecture Design - A requirements model and architecture design document was produced. This contained the web-app requirements model and web app architecture and design.

F. Test Cases and Documentation - A test case document containing the test scenario, test steps, and expected results was created. This was used as a reference for the inter-project and client testing in SE 2. The testing deliverable was the last deliverable for SE 1.

Software implementation and maintenance was covered in SE 2. Similarly, the students utilize the project blog to document their weekly progress. Below are the activities done by the students.

A. Software implementation - Weekly Progress Reports of the status of the project activities were provided. This contained the project burn-down chart, features completed and weekly progress details on completed and in progress activities. Materialized risks are also identified weekly.

B. Software Maintenance Plan - A software maintenance plan was also produced. Here the system description and maintenance procedures were provided. Details of the system description include: Point of Contact, Security, Computer Hardware, Supporting Software, and Personnel. Maintenance procedures starts off with conventions, followed by the maintenance procedures, verification procedures and error conditions.

C. Inter-project Test Evaluations - Each project team did test evaluations of three other projects. Here they provided the Software Testing Report as well as Maintainability Rating for the applications developed. A Mobile App Usability Checklist was utilized across the projects with results along with the test cases specific to the applications.

D. Project Evaluation by Client - The client was asked to evaluate the application. An evaluation form and acceptance criteria was used to document the client feedback. Project evaluations were done after the software is released. The students are required to do at least two software releases. As such software features were bundled and evaluated by the clients at least twice.

Some of the university A student projects mentioned the use of scrum and have even identified a scrum master in one of the projects. For this project, the team produced weekly progress reports where the features completed and overall project completion is detailed. The actual conduct of the scrum is not clear based on the documentations provided. There is no mention of sprints, sprints planning, daily scrum and sprint review. The same goes for another project which have identified risks of not completing a feature during a sprint.

3.2.3.2 UNIVERSITY B - WATERFALL AND AGILE-SCRUM

For University B, only final documents were provided to the researchers.

For SE 1 the project team 1 produced a (a) System Project Management Plan, (b) System Requirement Specifications, (c) System Design Document and (d) System Test Plan. These outputs are similar to the ones produced by University A. Note that for University B no prototype was created for both projects. The software application was produced as part of SE 2 requirement along with a project defense. For software maintenance, ITIL processes (Brenner, 2006) was mentioned in one of the projects though its use is not elaborated.

For project team 2, Agile-Scrum was used and documented as the project method in the final documents. Some of the reasons for this choice is the speed of the time to market and the advantage of close collaboration.

Requirements were stated as user stories and developments were done in sprints. Similar to student project A, a project schedule was also produced which reflected the development in sprints. System requirements were detailed (with the aid of use case diagrams, entity diagrams to name a few). Mock up of the application pages were also provided. This was followed by a detailed test plan. No mention of software maintenance was provided for this project.

For University B, students claimed (in the survey) that they use waterfall. 75% of the students claimed that they integrate testing throughout the product lifecycle and that they have a highly disciplined documentation process which is because of the required submission for the 2 semesters. However, when asked if their project development involved iteration, 75% of them also answered that they do have iterative process.

3.3 *Software engineering in the industry: Start-up companies*

3.3.1 *Start-up companies*

We interviewed representatives (i.e. Chief Technology officer and developers) from two start-up companies (X, Y) with head offices in Metro Manila. Both companies started with the founding members of six and have grown to 17-20 employees in a span of almost two years.

3.3.2 *Nature of the project and project teams*

Company X provides software development services (e.g. premier technology, game development, app development and end to end server for applications). They have a head office where employees can meet. Employees are also allowed to do remote work. The company projects are (1) client focused - client provides the requirements and (2) product/ problem focused - Product owner identifies product/problem that can be solved by technology. Projects are driven by a fixed set of tasks (i.e. features, functions) which resources can commit to finish within a week. Projects have a project manager and/or product manager, technical lead and business representative.

Company Y provides financial services focusing on platform creation for the Philippine market (e.g. rural banks). The projects are prototype focused and is composed of subject matter experts. The development team has a front end, back-end and integrator resources.

3.3.3 *Software engineering practices*

3.3.3.1 COMPANY X: WATERFALL, AGILE METHOD AND PROTOTYPING.

Company X shared two projects: one problem focused and another one client driven.

For the problem focused project, the project team used the agile method. It started off with a needs analysis with the project lead talking with the business. After which a product (e.g. a gamified marketing tool) was identified and designed by the lead designer. A prototype was done through Adobe XD with focus on the minimum viable product (MVP). The design was shown to the development team and client for funding approval. Two main developers: back-end and front-end proceeded with product development. Cross-developer testing is done. The application is then made available in the App store where employees are encouraged to use it. Feedback from users are captured in one to two days. These feedback are compiled, which focuses mainly on improvements on the user interface and user experience of things. There is minimal quality assurance. Updates of projects are provided in Trello and there is a weekly scrum. In terms of prioritization of features, the main functionality gets chosen first. At some point the project manager can change depending on the project focus: business or technical.

The client driven project, a fleet management system, utilized the agile method initially. The project team opted to provide a custom-built solution as the existing vendor solutions had too much features. A modularized solution, customized to business was chosen. For the first 6 months, the baseline features were developed with the backend ready to be connected and disconnected. Testing was done after the prototype (MVP) was developed. Then the developers discovered the missing features.

Afterwhich, the team went back to more planning, this time opting for the waterfall method. One week was spent for the concept and two more months for the features, user stories, development. At the time of the interview, the development team is doing more market research.

3.3.3.2 COMPANY Y: AGILE METHOD, PROTOTYPING AND TIMEBOXING

The project team is prototype focused and is event/goal driven. Once specific features for the proof of concept are identified, the developers build the prototype and delivers it the soonest possible time agreed. Development immediately commence with the identified use cases: basic features. Developers sit together and once specific features are done, they are integrated/connected and tested. The three developers do their testing. There is a daily scrum where the team discusses what has been done, if on time, how to proceed and blockers. Team members are independent and there is no micromanagement. Team does automated testing though at times developer prefers manual testing when automated is more effort extensive.

In terms of documentations, only wireframes are produced. Communication is through the following channels:

- Daily scrum - Only 2 mins online update,
- Chats though most time developers are not disturbed so they can focus on their work.
- Face to face discussions are triggered by events (e.g. breaking change).

Tools are used, with the development tools driven by the tech stack. Management tools include Slack which was switched to Rig-central, Trello and JIRA switched to Monday.com (list format).

The interviewee cited some challenges encountered in using the agile method as well as some solutions implemented.

- No intense planning and features are appended. Currently, the data per step are now planned out. For planning, the team uses timeboxing (2 days, else they risk taking too long in the ideation process).
- More ideal to have twice a week update rather than daily update. There are days when developers productivity drop (and minimal accomplishments). There are also days when the productivity is high and developers work 24 hours non-stop. The developers cope with the daily updates by reporting only a subset of their accomplishments (so they have something to report on other days).

The advantages of the method they use are: (1) fast turn-around time: use case and prototype can be presented early on with the end goal to test end to end and (2) motivation is higher .

3.4 Software engineering in the industry: Traditional companies

3.4.1 Traditional companies

The researchers also highlighted on two multinational IT companies in the Philippines (with employees >6000), hereon referred to as Company A and Company B. Both companies are global companies providing end-to-end solutions and are leaders in the software development industry.

3.4.2 Nature of the project and project teams

Company A is divided into technology that they support. The project teams are assembled accordingly with developers and project managers usually from the same technology group. The Philippine office of Company A is CMMI (Capability Maturity Model Integration) Level 5 certified and as such, waterfall is one of the methods used and prescribed. CMMI Levels are a measure of organization process maturity with Level 5 focused on process improvement and innovation (Curtis, 2019). There is also a software quality team which ensures that company policies, standards and templates are utilized. The group provides training of these processes across the company. Projects are audited regularly with best practices derived and shared across the groups.

For Company B, there is a project management office and a software quality management team. The project management office provides the project managers and scrum masters/practitioners with the developers spread across the organization. The software quality management team handles the CMMI certifications and audits the actual conduct of the project (including the documents). Different software models are utilized based on the nature of the project. For those which software requirements are standard, waterfall is used. Change management process is utilized to handle requirements changes. Agile-scrum is used for projects which requirements are known at a high level, client representatives have high availability, technology allows for it and fast feature delivery. Prototyping is utilized for exploratory projects.

3.4.3 *Software engineering practices*

Of the 15 respondents from this company 8 of them answered that they are using agile-scrum model while 3 responded it is a combination of scrum and waterfall. The other 4 claimed that they are using prototyping, Agile-XP, Spiral Model or Waterfall purely. Given the size of Company A, it is not surprising that they would be using different models for different teams. This is based on the requirement and nature of the team's project.

One of the best agile practice of Company A is following design/coding standards set by the management (Q17). 80% of the respondents answered ALWAYS to this question.

It seems that Company A has a stricter policy on SE practices that is why respondents in this group gave highest ranking on several of the given choices among them are:

- We integrate testing throughout the lifecycle, enabling regular inspection of the working product as it develops (Q2).
- We follow design/coding standards set by the management (Q17).
- Our project development is iterative (Q8).
- The goal is set at the start of development (Q7).
- The client identifies features of project to prioritize and deliver first (Q13).
- We have tools for automated testing (Q15).

All 15 respondents coming from Company B indicated that they are using Agile-Scrum in their projects. 8 of them responded Agile-Scrum only while the rest answered Scrum plus other model process like Waterfall Model, Prototyping and Agile-XP.

Although respondents in Company B agree that they are using Agile-Scrum as their company's model process, their responses in SE practices is more diverse. Many answered that such practice was not done ALWAYS but OFTENTIMES or SOMETIMES practiced. High responses for ALWAYS practiced are:

- We integrate testing throughout the lifecycle, enabling regular inspection of the working product as it develops (Q2).
- The goal is set at the start of development (Q7).
- Our team identify features of project to prioritize and deliver first (Q12).

4   CONCLUSION AND RECOMMENDATION

Waterfall is used in traditional/corporate and start-up companies to develop standard software for clients as confirmed in the survey and interview results. Because of the fast-paced and fast-changing characteristic of software development, agile process model, more specifically scrum, is the preferred SLDC model by 84% (66 out of 79) of all respondents who are developers/software engineers from different IT companies. In terms of the SE practices, both traditional and start-up companies rated high (73% and 62% for Traditional companies, 63% and 63% for Start-up companies, respectively) on two agile practices namely: (i) integrating testing throughout the lifecycle, enabling regular inspection of the working product as it develop and (ii) project development is iterative. Other agile practice were not always implemented (e.g. Q1- Strong Client involvement, Q3- Quick and frequent delivery of new features, Q12- The team will identifies the features of project to prioritize and deliver first). It can also be observed that software companies are actually incorporating other practices which are not scrum method into their process. Examples of this are Q13- The client identifies the features of project to prioritize and deliver first and Q17- Following a design/coding standard set by the management which of XP practices.

Some of the most common challenge in IT software companies include time-constraint and re-work due to requirement change or additional requirement. This leads to overtime and reduced productivity. In agile method, challenges mentioned is the lack of planning which is augmented by time-boxed planning as well. It was also suggested to change the scrum timings from daily to weekly. The developers cope with the daily updates by reporting only a subset of their accomplishments (so they have something to report on other days).

Waterfall and prototyping is the most common process model in the academe. 8/9 of the student projects used waterfall while 7/9 used prototyping. Agile scrum was mentioned in some projects of University A and was practiced in a University B project. In the student projects, the choice of the software engineering method is driven by the course requirements, including the resource and schedule constraints. We observe that waterfall is used by both universities. The use of document templates (e.g. project documentation, requirements modeling) help ensure that students provide the desired level of detail from an academic context. Project blog is useful in the weekly update and progress tracking. In some companies setup of the project repository is a standard practice. Most of the time, the weekly deliverables were provided. Improvement can be made in risk management(Boehm,1991), where weekly risk can further be identified. Once a risk is realized it becomes an issue. Here issue management(Bannerman,2008) can be applied. Prototyping is also practiced in University A as it was identified as a course requirement.

The requirement of the course syllabus of the university dictates the process model that students undertake. Addressing this concern could help familiarize the students in the real industry setting of software development.

In 2016, V. Garousi et. al. did a systematic literature review of the challenges and best practices in industry-academia collaborations in software engineering. In the Philippines, the academic community usually sends students to companies as part of the on-the-job training. Some students also work with companies as part of their SE project.

Academe-government collaborations in software engineering are richer with the researcher proposing projects and the government funding it. The Commission on Higher Education (CHED) and Department of Communications and Information Technology (DICT) are promoting more industry-academe collaboration such that research done is utilized or focused on industry needs. In line with this, a study on the current industry-academe collaboration projects can be a further area of research. Analysis of the software engineering practices from these projects, to identify the SE practices used, challenges and best practices can be an extension of this work.

REFERENCES

Alshamrani A. and Bahattab A. 2015. A Comparison Between Three SDLC Models Waterfall Model, Spiral Model, and Incremental/Iterative Model. *IJCSI International Journal of Computer Science Issues*, Volume 12, Issue 1, No 1, January 2015. p.106–111

Bannerman P.L. 2008. Risk and Risk Management in Software Projects: A Reassessment. J. *Systems and Software*, vol. 81, no. 12, pp. 2118–2133.

Boehm B. 1991. Software risk management: principles and practices. *IEEE Software*, 8, 1, 32–41.

Boehm B.W., Gray T.E., and Seewaldt T.1984. Prototyping versus specifying: a multiproject experiment. *IEEE transactions on Software Engineering 3* (1984): 290–303.

Brenner M. 2006. Classifying ITIL processes: A taxonomy under tool support aspects. *The First IEEE/ IFIP International Workshop*, pp. 19–28. IEEE Press.

Christensen L., Johnson R.B. & Turner L.A. 2011. *Research Methods, Design, and Analysis*. Boston MA: Pearson

Curtis B. 2019. Organizational Maturity: The Elephant Affecting Productivity. In: Sadowski C., Zimmemann T. (eds) Rethinking Productivity in Software Engineering. Apress, Berkeley, CA: 241–250.

DeVellis R. F. 1991. Scale Development Theory and Applications. *Applied Social Research Methods Se ries*, Vol. 26. Newbury Park, CA Sage Publications.

Extreme Programming (XP) vs Scrum. Retrieved from https://www.visual-paradigm.com/scrum/extreme-programming-vs-scrum/

Garousi O., Petersen K. and Özkan B. 2016. Challenges and Best Practices in Industry-academia Collaborations in Software Engineering. *Information and Software Technology* 79, C, 106–127. DOI:https://doi.org/10.1016/j. infsof.2016.07.006

Garousi V., Coşkunçay A., Can A. B., and Demirörs O. 2015. A Survey of Software Engineering Practices in Turkey. *Journal of Systems and Software*, vol. 108, pp. 148–177.

IEEE Standard Glossary of Software Engineering Terminology, IEEE std 610.12-1990, 1990.

Jalote P., et al.2004. Timeboxing: A Process Model for Iterative Software Development, Journal of Systems and Software,vol. 70, issue 2, 2004, pp. 117–127.

Raval R and Rathod H, 2013. Comparative Study of Various Process Model in Software Development, *International Journal of Computer Applications* (0975 – 8887), Vol 82 – No.18, November 2013.

Sandberg, A. B., & Crnkovic, I. 2017. Meeting industry: academia research collaboration challenges with agile methodologies. In *Proceedings of the 39th International Conference on Software Engineering: Software Engineering in Practice Track* (pp. 73–82). IEEE.

Sison R and Yang T. 2007. Use of Agile Methods and Practices in the Philippines. In: *14th Asia Pacific of Software Engineering Conference*, APSEC.

Sison R., Jarzabek S., Hock O.S., Rivepiboon W., and Hai N.N. 2006. Software practices in five ASEAN countries: an exploratory study, *Proceedings of the 28th international conference on Software engineering* Shanghai, China: ACM.

Software Development. IT-BPM Roadmap 2022: Accelerate PH, Future Ready. Retrieved from: http://www.boi.gov.ph/wp-content/uploads/2018/02/Software-Development-December-2017.pdf

Zelkowitz M., Yeh R., Hamlet R., Gannon J., and Basili V. 1984. Software engineering practices in the U.S. and Japan. *IEEE Comput.* 17, 6 (June 1984), 57–66.

Theory and Practice of Computation – Nishizaki et al (eds)
© 2021 Taylor & Francis Group, London, ISBN 978-0-367-41473-3

Efficiency of reduction methods on the size and order of workflow net reachability graphs

Louis Anthony Agong & H.N. Adorna
University of the Philippines, Diliman, Philippines

ABSTRACT: The paper presents the effect of the reduction rules on the size of the resulting workflow net as well as the size and order of its reachability graph. The proposed method involves looking for the nodes and arcs in the reachability graph that is affected by the removal of places and transition of the Petri net. We determined that the reduction rules, with the exception of one rule, applied to the Petri net provides a reduction on the size and order of the its corresponding reachability graph.

1 INTRODUCTION

A Petri net (PN) (Petri, 1962) is a mathematical tool that can be used in describing concurrent systems. Petri net properties such as liveness and boundedness are used to determine if the system can run indefinitely while using limited amount of resources. However, other PN properties such as the reachability of its markings are difficult to compute. The reachability problem for Petri net markings is proven to be EXPSPACE-hard (Lipton, 1976). Because of this it is necessary to reduce a complex net to a more simpler and smaller one. Reduction methods on Petri nets are employed to produce a more abstracted model while preserving the property that we want to analyze. Inversely, synthesis rules allow us to produce a more refined model.

Analysis of Petri nets could be done on its reachability graph. The construction of a reachability graph involves the enumeration of all possible reachable markings from the initial marking. However, due to state-space explosion the size of the reachability graph increases drastically as the size of the Petri net increases. The initial marking is also a factor in the size of the reachability graph. In this paper we are interested in the effects of applying the reduction methods in the reachability graph of a workflow net. Specifically we are observing the effects of Bride's (Bride et al., 2017) reduction rules on its effect on the resulting reduced workflow net as well as the reachability graph on the workflow nets. These reduction rules are built on top of the results obtained from Murata (Murata, 1989), and Desel and Esparza (Desel and Esparza, 1995). For the reachability graph analysis we will mainly look around the nodes and arcs directly affected by the removal of nodes from the source workflow net.

The paper is organized as follows. In Section 2 we present the definitions and preliminary concepts that will be used in our analysis. Section 3 details our method of analyzing the efficiency of Bride's reduction rules on the size and order of the reachability graph of the workflow net. Section 4 concludes the paper and presets possible future work.

2 PRELIMINARIES

In this section we provide the definitions used for analysis of efficiency. In particular, we will define Petri nets, its properties, and definitions for our measure of efficiency.

Definition 1.1 (Petri Net) *A Petri Net is a 3-tuple, (P, T, F) where:*
*– $P = \{p_1, p_2, \ldots, p_m\}$ is a finite set of **places***

- $T = \{t_1, t_2, \ldots, t_m\}$ *is a finite set of* **transitions**
- $F \subseteq (P \times T) \cup (T \times P)$ *is a set of arcs*
- *In particular, a place p is called an* **input (output)** *place of a transition t if* $(p, t) \in F$
  $((t, p) \in F)$ *and a transition t is called an* **input (output)** *transition of a place p if*
  $(t, p) \in F$ $((p, t) \in F)$
- *We define a* **marking** *as a mapping* $P \to N$ *that represents the number of tokens on the places.*
  If p is a place, p^k represents k tokens on p. Thus, $M = p_1^{k_1} p_2^{k_2} \cdots p_n^{k_n}$ represents a marking
  that has k_i tokens in place p_i for $i = 1, 2, \ldots, n$.
- $\bullet t = \{p | (p, t) \in F\}$ *= the set of input places of t*
- $t^\bullet = \{p | (t, p) \in F\}$ *= the set of output places of t*
- $\bullet p = \{t | (t, p) \in F\}$ *= the set of input transitions of p*
- $p^\bullet = \{t | (p, t) \in F\}$ *= the set of output transitions of p*
- *If $G \subseteq P \cup T$ then $G^\bullet = \cup_{g \in G} g^\bullet$*
- *If $G \subseteq P \cup T$ then $\bullet G = \cup_{g \in G} \bullet g$*

Definition 1.2 (Transition Firing Rules) *A transition t is said to be* **enabled** *if each input place
p of t contains at least one token. An enabled transition may or may not fire. If a transition t*
fires, *then t consumes one token from each input place p of t and produces one token for each
output place of t. Let M_a, M_b be two markings. We denote $M_a \xrightarrow{t} M_b$ the fact that transition t
is enabled by marking M_a and firing t results in the marking M_b. Let $M_1, M_2, \ldots M_n$ be mark-
ings and $\sigma = t_1 t_2 \ldots t_{n-1}$ as sequence of transitions. We denote $t' \in \sigma$ if $t' \in \{t_1, \ldots, t_{n-1}\}$. We
denote $M_1 \xrightarrow{\sigma} M_n$ the fact that $M_1 \xrightarrow{t_1} M_2 \xrightarrow{t_2} \cdots \xrightarrow{t_{n-1}} M_n$.*

Definition 1.3 (Reachability) *Let M be a marking of a PN N. A marking M' is said to be*
reachable *from M if there exists a sequence of firing transitions that transforms M to M'.
Denote $R(N, M)$ as the set of all reachable markings from M.*

Definition 1.4 (Reachability Graph) *Let $N = (P, T, F)$ be a Petri net initialized by marking
M_i, denoted by (N, M_i). A* **reachability graph** *RG of $R(N, M_i)$ is a directed graph where the
nodes are the markings reachable from M_i. An ordered pair (M_x, M_y) is an arc of RG if there
exist a transition $t \in T$ such that $M_x \xrightarrow{t} M_y$. The arc (M_x, M_y) is labeld by t.*

Definition 1.5 (Workflow Net) *(van der Aalst, 1996) A petri net N is a* **workflow net** *if*

- *N has a source place i such that $\bullet i = \emptyset$*
- *N has a sink place o such that $o^\bullet = \emptyset$*
- *If we add a transition t' to N which connects place o with i (i.e. $\bullet t' = \{o\}$ and $t'^\bullet = \{i\}$), then
  the resulting Petri net is strongly connected.*

Definition 1.6 (k-soundness) *Let $N = (P, T, F)$ be a workflow net with source place i and
a sink place o. Denote i^k and o^k to be the marking with k tokens in i and o, respectively. N is said
to be k-**sound** if*

- $\forall M \in R(N, i^k), o^k \in R(N, M)$
- $\forall t \in T, \exists M \in R(N, i^k)$ *such that t is enabled*

Definition 1.7 (Generalised Soundess) *A workflow net N is* **generalised sound** *if $\forall k \in N$, N is
k-sound.*

Definition 1.8 (Size and Order of the Reachability graph) *Let N be a Petri net and M_i be
a marking. Let $RG = (V, E)$ be the reachability graph of $R(N, M_i)$ where V is the set of nodes
and E is the set of arcs.*

- *The* **order** *of RG, denoted order(RG), is $|V|$*
- *The* **size** *of RG, size(RG), is $|E|$*

We will use *order* and *size* as our measure for determining efficiency of the reduction rules.
For example, as illustrated in Figure 1, we have a workflow net N and a reachability graph
RG of $R(N, i^3)$ where $order(RG) = 10$ and $size(RG) = 18$.

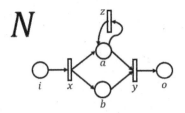

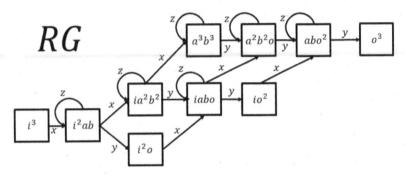

Figure 1. Reachability graph RG of $R(N, i^3)$.

3 EFFICIENCY OF THE REDUCTION RULES

In this section, we show the effect of the Bride's reduction rules on the size of the resulting work-flow net as well as the size and order of its reachability graph. The reduction rules are proven to strongly preserve generalised soundess on the class of workflow nets. For the preservation proof, discussion and examples on the reduction rules we refer you the source paper (Bride et al., 2017). We only list the conditions and constructions for each rule. We also assume that the workflow net is bounded or else we will be dealing with a reachability graph of infinite size and order. For the accompanying figures (Figures 2,3,4,5,6), dashed nodes and arcs represent the removed elements of the reachability graph while heavy lines represent the added elements.

3.1 R_1: Remove Place ($\phi_{RemoveP}$)

Conditions on N:

$$\exists p \in P \backslash \{i, o\} \tag{1}$$

$$\exists G = \{g_1, \ldots, g_n\} \subseteq P \backslash \{i, o\} \tag{2}$$

$$p^\bullet = G^\bullet \tag{3}$$

$$^\bullet p = {}^\bullet G \tag{4}$$

$$\forall r, s \in G, r \neq s \Rightarrow {}^\bullet r \cap s^\bullet = \emptyset = r^\bullet \cap s^\bullet \tag{5}$$

Construction of $\overline{N}$:

$$\overline{P} := P \backslash \{p\} \tag{6}$$

$$\overline{T} := T \tag{7}$$

$$\overline{F} := F \backslash (((\{p\} \times p^\bullet) \cup ({}^\bullet p \times \{p\})) \tag{8}$$

67

In other words, the construction of $\overline{N}$ only removes p and all arcs to and from p.

Lemma 1.1 *Let $f : R(\overline{N}, i^k) \to R(N, i^k)$ such that $f(\overline{M}) = M$ where $M(q) = \overline{M}(q)$ for all $q \in \overline{P}$ and $M(p) = \overline{M}(g_1) + \overline{M}(g_2) + \cdots + \overline{M}(g_n)$. Then f is a bijection.*

Proof. Let $N = (P, T, F)$ and $\overline{N} = (\overline{P}, \overline{T}, \overline{F})$ be workflow nets based on the conditions and construction in 3.1.

Injection: Let $\overline{M}_a, \overline{M}_b \in R(\overline{N}, i^k)$ such that $f(\overline{M}_a) = f(\overline{M}_b)$. Then, from the definition of f, $\overline{M}_a(q) = f(\overline{M}_a)(q) = f(\overline{M}_b)(q) = \overline{M}_b(q)$ for all $q \in \overline{P}$ which concludes that f is an injection.

Surjection: Let $M \in R(N, i^k)$.

Take $\overline{M}$ where $\overline{M}(q) = M(q)$ for all $q \in P \backslash \{p\}$. It can be proven that $M(p) = M(g_1) + \cdots + M(g_n) = \overline{M}(g_1) + \cdots + \overline{M}(g_n)$ so that, $f(\overline{M}) = M$ which proves surjection and therefore f is a bijection.

Observe that

1. $|\overline{P}| = |P| - 1$
2. $|\overline{T}| = |T|$
3. $|\overline{F}| = |F| - |p^\bullet| - |{}^\bullet p|$

Proposition 1 *Let $\overline{N}$ be the net constructed after applying $\phi_{RemoveP}$ in N. Let RG and $\overline{RG}$ be the reachability graph of $R(N, i^k)$ and $R(\overline{N}, i^k)$, respectively. Then,*

1. $order(\overline{RG}) = order(RG)$
2. $size(\overline{RG}) = size(RG)$

Proof. It follows from Lemma 3.1 that we can obtain a bijective mapping from M to $\overline{M}$. This shows that we can obtain an isomorphism between the reachability graph of N and the reachability graph of $\overline{N}$. The removal of p doesn't change the structure, order, and size of the resulting reachability graph.

3.2 R_2: Remove Transition ($\phi_{RemoveT}$)

Let D be a set of places. Define the function $\nu : D \to (P \to N)$ such that

$$\forall d \in P, \nu(D)(d) = \begin{pmatrix} 1 & \text{if } d \in D \\ 0 & \text{otherwise} \end{pmatrix}$$

Conditions on N:

$$\exists t \in T \tag{9}$$

$$\exists G = \{g_1, \ldots, g_n\} \subseteq T \backslash \{t\} \tag{10}$$

Define $\nu_t := \nu(t^\bullet) - \nu({}^\bullet t)$ and $\nu_G := \nu(g_1^\bullet) + \cdots + \nu(g_n^\bullet) - \nu({}^\bullet g_1) - \cdots - \nu({}^\bullet g_n)$

$$\nu_t = \nu_G \tag{11}$$

$$\forall r, s \in G, r \neq s \Rightarrow {}^\bullet r \cap {}^\bullet s = \emptyset = r^\bullet \cap s^\bullet \tag{12}$$

$$(\exists t_s \in T \backslash (\{t\} \cup G), \forall g \in G, {}^\bullet g \subseteq t_s^\bullet)) \text{or}(|G| = 1) \tag{13}$$

Let $outArc := \{t\} \times t^\bullet, outArc := {}^\bullet t \times \{t\}$.
Construction of $\overline{N}$:

$$\overline{P} := P \tag{14}$$

$$\overline{T} := T \backslash \{t\} \tag{15}$$

$$\overline{F} := F \backslash (inArc \cup outArc) \tag{16}$$

Lemma 1.2 *Consider the conditions in N from 3.2. Let M_i, M_j be markings of N. Then,*
$M_i \xrightarrow{t} M_j$ *if and only if* $M_i \xrightarrow{\sigma} M_j$ *for some transition sequence σ such that $t \notin \sigma$ and for all*
$g \in \sigma, g \in G$ *and* $g_i, g_j \in \sigma \Rightarrow g_i \neq g_j$.

Proof. Note that from the condition above (11), $^\bullet t = \cup_{g \in G}(^\bullet g)$ and $t^\bullet = \cup_{g \in G}(g^\bullet)$

Suppose $M_i \xrightarrow{t} M_j$ in N. Since M_i enables t, each place $p \in {}^\bullet t$ has at least one token. This implies that each transition $g \in G$ is enabled. Also, due to ((12)) firing a place g_i does not disable other g_j in G. Construct $\sigma = g_1 g_2 \ldots g_n$ so that $M_i \xrightarrow{\sigma} M'$. Firing each $g \in G$ produces a token in $g^\bullet$. Due to (12), each firing in σ only produce exactly one token on $g^\bullet$ for all $g \in G$. Since $t^\bullet = \cup_{g \in G}(g^\bullet)$, we have $M' = M_j$.

Suppose $M_i \xrightarrow{\sigma} M_j$ in $\overline{N}$ for some transition sequence σ such that $t \notin \sigma$ and for all $g \in \sigma$, $g \in G$ and $g_i, g_j \in \sigma \Rightarrow g_i \neq g_j$. Then each transition $g \in G$ is enabled which implies that t is enabled. Firing t will produce a marking M' in N. Each firing will produce a token in each place in $g^\bullet$ for all $g \in G$. Due to (12) exactly one token is placed. This means that, due to (11), one token is placed for all place in $t^\bullet$. Therefore $M' = M_j$

Observe that

1. $|\overline{P}| = |P|$
2. $|\overline{T}| = |T| - 1$
3. $|\overline{F}| = |F| - |t^\bullet| - |{}^\bullet t|$

Proposition 2 *Let $\overline{N}$ be the net constructed after applying $\phi_{RemoveT}$ in N. Let RG and $\overline{RG}$ be the reachability graph of $R(N, i^k)$ and $R(\overline{N}, i^k)$, respectively. Then,*

1. $order(\overline{RG}) = order(RG)$
2. $size(\overline{RG}) = size(RG) - |\{(M_i, M_j) : M_i \xrightarrow{t} M_j\}|$

Proof. For (1), markings affected by the removal, in this case M_i, M_j such that $M_i \xrightarrow{t} M_j$, are not removed since we can find a sequence of transition firing σ such that $M_i \xrightarrow{\sigma} M_j$.

For (2), only edges in R labelled by t are removed.

The effect of Proposition 2 is illustrated in Figure 2.

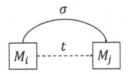

Figure 2.   Effect of $\phi_{RemoveT}$ on the reachability graph.

3.3 R_3: Remove Self-loop ($\phi_{RemoveST}$)

Conditions on N:

$$\exists t \in T \tag{17}$$

$$^\bullet t = t^\bullet \tag{18}$$

$$\exists t_s \in T \setminus \{t\} \text{ such that } {}^\bullet t \subseteq t_s^\bullet \text{ or } t^\bullet \subseteq {}^\bullet t_s \tag{19}$$

Construction of $\overline{N}$:

$$\overline{P} := P \tag{20}$$

$$\overline{T} := T \setminus \{t\} \tag{21}$$

$$\overline{F} := F \setminus (((\{t\} \times t^\bullet) \cup ({}^\bullet t \times \{t\})) \tag{22}$$

Observe that

1. $|\overline{P}| = |P|$
2. $|\overline{T}| = |T| - 1$
3. $|\overline{F}| = |F| - |t^\bullet|$

Proposition 3 *Let $\overline{N}$ be the net constructed after applying $\phi_{RemoveST}$ in N. Let RG and $\overline{RG}$ be the reachability graph of $R(N, i^k)$ and $R(\overline{N}, i^k)$, respectively. Then,*

1. $order(\overline{RG}) = order(RG)$
2. $size(\overline{RG}) = size(RG) - |\{M : M \in R(N, i^k) \text{ and } M \text{ enables } t\}|$

Proof. For (2), if a marking $M \in R(N, i^k)$ enables t, then we have $M \xrightarrow{t} M$. These are self loops labeled by t. Construction of $\overline{N}$: removes these self loops and the number of self loops labeled by t are determined by the number of markings that enables t.

For (1), the number of markings are unaffected since only self loops are removed.

The effect of Proposition 3 is illustrated in Figure 3.

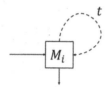

Figure 3.   Effect of $\phi_{RemoveST}$ on the reachability graph.

3.4 R_4: Remove Transition-Place $(\phi_{RemoveTP})$

Conditions on N:

$$\exists p \in P\backslash\{i, o\} \tag{23}$$

$$^\bullet p = \{t\} \tag{24}$$

$$t^\bullet \neq \{p\} \Rightarrow \forall ot \in p^\bullet, {}^\bullet ot = \{p\} \text{ and } t^\bullet \cap ot^\bullet = \emptyset \tag{25}$$

$$t^\bullet = \{p\} \Rightarrow \forall ot \in p^\bullet, {}^\bullet t \cap {}^\bullet ot = \emptyset \text{ and } (\exists ot \in p^\bullet, {}^\bullet ot = \{p\} \text{ or } \forall ip \in {}^\bullet t, ip^\bullet = \{t\}) \tag{26}$$

Let
$OT = p^\bullet, IP = {}^\bullet t, OP = t^\bullet\backslash\{p\}, outT := \{t\} \times OP, inT := IP \times \{t\}, outP :-$
$= \{p\} \times OT, inArc := IT \times OP, outArc := OT \times OP.$
Construction of $\overline{N}$:

$$\overline{P} := P\backslash\{p\} \tag{27}$$

$$\overline{T} := T\backslash\{t\} \tag{28}$$

$$\overline{F} := (F \cup inArc \cup outArc)\backslash((t, p) \cup inT \cup outT \cup outP) \tag{29}$$

Lemma 1.3 *Let M_i, M_j, M_k be markings in N such that $M_i(p) = 0$. Let $ot \in OT$. Then, $M_i \xrightarrow{t} M_j \xrightarrow{ot} M_k$ in N if and only if $M_i \xrightarrow{ot} M_k$ in $\overline{N}$.*
Proof. Suppose $M_i \xrightarrow{t} M_j \xrightarrow{ot} M_k$ in N.

Suppose, in N, that $t^\bullet \neq \{p\}$ and for all $ot \in OT$, ${}^\bullet ot = \{p\}$ and $t^\bullet \cap ot^\bullet = \emptyset$. M_i enables t in N so we can obtain M_j by firing t, thus consuming tokens in ${}^\bullet t$ and producing in $t^\bullet$ including

70

p. We can then obtain M_k by firing ot, consuming a token in p and producing tokens on each place of $ot^\bullet$. Thus,

$$M_k(q) = \begin{cases} M_i(q) - 1, & \text{if } q \in IP, \\ M_i(q) + 1, & \text{if } q \in OP, \\ M_i(q) + 1, & \text{if } q \in (t^\bullet \backslash \{p\}), \\ M_i(q), & \text{otherwise.} \end{cases}$$

Note that from the construction of $\overline{N}$, we have $IP \subseteq {}^\bullet ot$ and $ot^\bullet = OP \cup (t^\bullet \backslash \{p\})$ for all $ot \in OT$. We can also obtain M_k from M_i by firing ot since firing ot will consume tokens ${}^\bullet ot \cup IP$ and produce tokens on $OP \cup (t^\bullet \backslash \{p\})$.

Now, suppose condition (25) is true in N. M_i enables some $t \in IT$ in N so we can obtain M_j by firing t, thus consuming tokens in ${}^\bullet t$ and putting tokens in p and output places of $it^\bullet$. We can then obtain M_k by firing ot, consuming a token in $\{p\} \cup {}^\bullet ot$ and producing tokens on $ot^\bullet$. We have M_k such that

$$M_k(q) = \begin{cases} M_i(q) - 1, & \text{if } q \in IP, \\ M_i(q) - 1, & \text{if } q \in {}^\bullet ot, \\ M_i(q) + 1, & \text{if } q \in ot^\bullet, \\ M_i(q), & \text{otherwise.} \end{cases}$$

Note that from the construction of $\overline{N}$, we have $IP \subseteq {}^\bullet ot$ for all $ot \in OT$. We can also obtain M_k from M_i by firing ot since firing ot will consume tokens in $IP \cup {}^\bullet ot$ and produce tokens on $ot^\bullet$.
Therefore $M_i \xrightarrow{ot} M_k$ in $\overline{N}$.
We could use a similar argument for the converse.
Observe that

1. $|\overline{P}| = |P| - 1$
2. $|\overline{T}| = |T| - 1$
3. $|\overline{F}| = |F| + (|p^\bullet| - 1)(|{}^\bullet t||t^\bullet|) - 2 \cdot |p^\bullet|$

Proposition 4 *Let $\overline{N}$ be the net constructed after applying $\phi_{RemoveTP}$ in N. Let RG and $\overline{RG}$ be the reachability graph of $R(N, i^k)$ and $R(\overline{N}, i^k)$, respectively. Then,*

1. $order(\overline{RG}) = order(RG) - |\{M : M \in R(N, i^k) \text{ and } M(p) > 0\}|$
2. $size(\overline{RG}) = size(RG) - |\{(M_x, M_y) : M_x \xrightarrow{t} M_y\}| - |\{(M_x, M_y) : \exists t' \in T \backslash t, (M_x(p) > 0 \text{ or } M_y(p) > 0)$ where $M_x \xrightarrow{t} M_y\}| + |\{(M_x, M_y) : M_x(p) = M_y(p) = 0$ and $\exists t' \in T \backslash \{t\}, \exists M', M_x \xrightarrow{t} M' \xrightarrow{t'} M_y\}|$

Proof. For (1), markings that have tokens on p won't be reachable in $\overline{N}$.

For (2), first, all arcs involving t in the reachability graph will be removed. Second, since all markings that has tokens on p are removed, all transitions involving those markings will also be removed. However, due to Lemma 1.3 we can obtain new arcs (M_x, M_y) labeled ot where $M_x(p) = M_y(p) = 0$ and $M_x \xrightarrow{t} M' \xrightarrow{ot} M_y$ for some $ot \in T \backslash \{t\}$ in N so that $M_x \xrightarrow{ot} M_y$ in $\overline{N}$.

The effect of Proposition 4 is illustrated in Figure 4.

3.5 R_5: Remove Place-Transition ($\phi_{RemovePT}$)

Conditions on N:

$$\exists p \in P \backslash \{i, o\} \tag{30}$$

$$ {}^\bullet p = \{t\} \tag{31}$$

$$ {}^\bullet t \neq \{p\} \Rightarrow \forall it \in {}^\bullet p, (it^\bullet = \{p\} \text{ and } {}^\bullet it \cap {}^\bullet t = \emptyset \text{ and } ({}^\bullet it)^\bullet = \{it\}) \tag{32}$$

$$ {}^\bullet t = \{p\} \Rightarrow \forall it \in IT, t^\bullet \cap it^\bullet = \emptyset \tag{33}$$

71

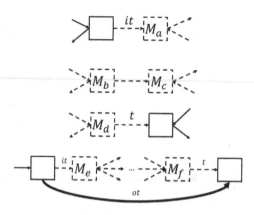

Figure 4. Effect of $\phi_{RemoveTP}$ on the reachability graph. Note that the labeled markings M_i have at least one token on p.

Let
$$OP := t^\bullet,\ IP:=^\bullet t\setminus\{p\},\ IT:=^\bullet p,\ outT := \{t\} \times t^\bullet,\ inT:= (^\bullet t\setminus p) \times t^\bullet,\ inP:=^\bullet p \times \{p\},\ inArc:= (^\bullet t\setminus\{p\})\times^\bullet p,\ outArc:=^\bullet p \times t^\bullet.$$
Construction of $\overline{N}$:

$$\overline{P} = P\setminus\{p\} \tag{34}$$

$$\overline{T} = T\setminus\{t\} \tag{35}$$

$$\overline{F} = (F \cup inArc \cup outArc)\setminus((p,t) \cup inT \cup outT \cup inP) \tag{36}$$

Lemma 1.4 Let M_i, M_j, M_k be markings in N such that $M_i(p) = 0$. Let $it \in IT$. Then, $M_i \xrightarrow{it} M_j \xrightarrow{t} M_k$ in N if and only if $M_i \xrightarrow{it} M_k$ in $\overline{N}$.

Proof. Suppose $M_i \xrightarrow{it} M_j \xrightarrow{t} M_k$ in N.

Suppose, in N, that $^\bullet t \neq \{p\}$ and for all $it \in IT$, $it^\bullet = \{p\}$, $^\bullet it \cap ^\bullet t = \emptyset$ and $(^\bullet it)^\bullet = \{it\}$. M_i enables some $it \in IT$ in N so we can obtain M_j by firing it, thus consuming tokens in $^\bullet it$ and putting a token in p. We can then obtain M_k by firing t, consuming a token in p and each place in IP and producing tokens on each place of OP. We have M_k such that

$$M_k(q) = \begin{cases} M_i(q) - 1, & \text{if } q \in {}^\bullet it, \\ M_i(q) - 1, & \text{if } q \in IP, \\ M_i(q) + 1, & \text{if } q \in OP, \\ M_i(q), & \text{otherwise.} \end{cases}$$

Note that from the construction of $\overline{N}$, we have $it^\bullet = OP$ and $IP \subseteq ^\bullet it$ for all $it \in IT$. We can obtain M_k from M_i in $\overline{N}$ by firing it since firing it will consume tokens in $IP \cup ^\bullet it$ and produce tokens on OP.

Now, $^\bullet t = \{p\}$ and for all $it \in IT$, $t^\bullet \cap it^\bullet = \emptyset$. M_i enables some $it \in IT$ in N so we can obtain M_j by firing it, thus consuming tokens in $^\bullet it$ and putting tokens in p and output places of $it^\bullet$. We can then obtain M_k by firing t, consuming a token in p and producing tokens on each place of OP. We have M_k such that

$$M_k(q) = \begin{cases} M_i(q) - 1, & \text{if } q \in {}^\bullet it, \\ M_i(q) + 1, & \text{if } q \in OP, \\ M_i(q) + 1, & \text{if } q \in it^\bullet \setminus OP, \\ M_i(q), & \text{otherwise.} \end{cases}$$

From the construction of $\overline{N}$ we can also obtain M_k from M_i by firing it since firing it will consume tokens in ${}^\bullet it$ and produce tokens on $OP \cup it^\bullet$.

Therefore $M_i \xrightarrow{it} M_k$ in $\overline{N}$.

We could use a similar argument for the converse.

Observe that

1. $|\overline{P}| = |P| - 1$
2. $|\overline{T}| = |T| - 1$
3. $|\overline{F}| = |F| + (|{}^\bullet p| - 1)(|{}^\bullet t||t^\bullet|) - 2 \cdot |{}^\bullet p|$

Proposition 5 *Let $\overline{N}$ be the net constructed after applying $\phi_{RemovePT}$ in N. Let RG and $\overline{RG}$ be the reachability graph of $R(N, i^k)$ and $R(\overline{N}, i^k)$, respectively. Then,*

1. $order(\overline{RG}) = order(RG) - |\{M : M \in R(N, i^k) \text{ and } M(p) > 0\}|$

2. $size(\overline{RG}) = size(RG) - |\{(M_x, M_y) : M_x \xrightarrow{t} M_y\}| - |\{(M_x, M_y) : \exists t' \in T \setminus t, (M_x(p) > 0 \text{ or } M_y(p) > 0) \text{ where } M_x \xrightarrow{t'} M_y\}| + |\{(M_x, M_y) : M_x(p) = M_y(p) = 0 \text{ and } \exists it \in IT, \exists M', M_x \xrightarrow{it} M' \xrightarrow{t} M_y\}|$

Proof. For (1), markings that have tokens on p won't be reachable in $\overline{N}$.

For (2), first, all arcs involving t in the reachability graph will be removed. Second, since all markings that has tokens on p are removed, all transitions involving those markings will also be removed. However, due to Lemma 1.4 we can obtain new arcs (M_x, M_y) labeled it where $M_x(p) = M_y(p) = 0$ and $M_x \xrightarrow{it} M' \xrightarrow{t} M_y$ for some $it \in IT \setminus \{t\}$ in N so that $M_x \xrightarrow{it} M_y$ in $\overline{N}$.

The effect of Proposition 5 is illustrated in Figure 5.

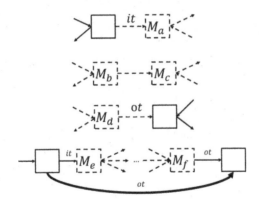

Figure 5. Effect of $\phi_{RemovePT}$ on the reachability graph. Note that the labeled markings M_i have at least one token on p.

3.6 R_6: Remove Ring ($\phi_{RemoveR}$)

Conditions on N:

$$\exists RP := \{p_1, \ldots, p_n\} \subseteq P \tag{37}$$

$$\exists RT := \{t_1, \ldots, t_n\} \subseteq T \tag{38}$$

$$\forall t \in RT, |{}^{\bullet}t| = |t^{\bullet}| = 1 \tag{39}$$

$$\forall p_i, p_j \in RP \text{ such that } p_i \neq p_j, {}^{\bullet}p_i \cap {}^{\bullet}p_j = p_i^{\bullet} \cap p_j^{\bullet} = \emptyset \tag{40}$$

$$\forall p_i, p_j \in RP, \exists \text{ a path } l \text{ from } p_i \text{ to } p_j \text{ such that all vertices of } l \text{ are in } RP \text{ and } RT \tag{41}$$

Let $ringArc := ((RP \times RT) \cup (RT \times RP)) \cap F, IT := {}^{\bullet}p_1 \cup \cdots \cup {}^{\bullet}p_n, OT := p_1^{\bullet} \cup \cdots \cup p_n^{\bullet}, removedArc := ((IT \times RP) \cup (RP \times OP)) \cap F, \ addArc := (IT \times \{p\}) \cup (\{p\} \times OP)$.

Construction of $\overline{N}$:

$$\overline{P} := (P \cup \{p\}) \setminus RP \tag{42}$$

$$\overline{T} := T \setminus RT \tag{43}$$

$$\overline{F} := (F \cup addArc) \setminus removedArc \tag{44}$$

Lemma 1.5 Let M_i, M_j, M_k, M_l, M' be markings in N such that $M_i(p) = 0$ for all $p \in RP$. Let $it \in IT$, $ot \in OT$ and σ be sequence of transitions such that $t \in \sigma$ for all $t \in RT$. Then $M_i \xrightarrow{it} M_j \xrightarrow{\sigma} M_k \xrightarrow{ot} M_l$ in N if and only if $M_i \xrightarrow{it} M' \xrightarrow{ot} M_l$ in $\overline{N}$.
Proof. Suppose $M_i \xrightarrow{it} M_j \xrightarrow{\sigma} M_k \xrightarrow{ot} M_l$. We have

$$M_l(q) = \left(\begin{array}{l} M_i(q) - 1, \text{ if } q \in {}^{\bullet}it, \\ M_i(q) + 1, \text{ if } q \in op^{\bullet}, \\ M_i(q), \text{ otherwise.} \end{array} \right.$$

Due to the construction of $\overline{N}$, we can obtain M' from M_i in $\overline{N}$ by firing it since firing it will consume tokens in ${}^{\bullet}it$ and produce tokens on p. Firing ot then consumes token on p and produces a token in $ot^{\bullet}$ producing the same marking M_l.
Suppose $M_i \xrightarrow{it} M' \xrightarrow{ot} M_l$. We have

$$M_l(q) = \left(\begin{array}{l} M_k(q) - 1, \text{ if } q \in {}^{\bullet}it, \\ M_k(q) + 1, \text{ if } q \in op^{\bullet}, \\ M(q), \text{ otherwise.} \end{array} \right.$$

Due to the conditions of N, we can M_j from M_i in N by firing it since firing it will consume tokens in ${}^{\bullet}it$ and produce tokens on some $p \in RP$. We can then obtain a σ that transfers the token to some $p' \in {}^{\bullet}ot$. Then, ot fires and consumes token on p' and produces a tokens in $ot^{\bullet}$ producing the same marking M_l.
Observe that

1. $|\overline{P}| = |P| - n + 1$
2. $|\overline{T}| = |T| - m$
3. $|\overline{F}| = |F| - 2m$

Proposition 6 *Define a function* $\mu : R(N, i^k) \to R(\overline{N}, i^k)$ *where if* $\mu(M) = M'$ *we have*

$$M'(q) = \left(\begin{array}{l} M(p_1) + \cdots + M(p_n), \text{ if } q = p, \\ M(q), \text{ otherwise.} \end{array} \right.$$

Let $\overline{N}$ *be the net constructed after applying* $\phi_{RemoveR}$ *in* N. *Let* RG *and* $\overline{RG}$ *be the reachability graph of* $R(N, i^k)$ *and* $R(\overline{N}, i^k)$, *respectively. Then,*

1. $order(\overline{RG}) = order(RG) - |\{M : M \in R(N, i^k) \text{ and } \exists p' \in RP, M(p') > 0\}|$

$+ |\{M : M \in \mu(R(N, i^k)) \text{ and } M(p) > 0\}|$

2. $size(\overline{RG}) = size(RG) - |\{(M_x, M_y) : M_x \xrightarrow{t} M_y, t \in RT\}|$

Proof. For (1), markings that have tokens on some $p \in RP$ won't be reachable in $\overline{N}$ and therefore will be removed. All places in the ring will be collapsed into a single place p. New markings are created where all tokens from each place in the ring will be transferred on p.

For (2), first, all arcs $t \in RT$ in the reachability graph will be removed. Second, all markings that has tokens on some place $rp \in RP$ are replaced by markings with tokens on p. Lastly, arcs that represent firings that add/remove tokens to/from any $rp \in RP$ are replaced by firings that add/remove tokens to/from p.

The effect of Proposition 6 is illustrated in Figure 6.

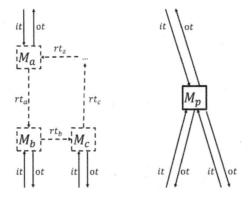

Figure 6. Effect of $\phi_{RemoveR}$ on the reachability graph. Note that the labeled markings M_i have at least one token on some $rp \in RP$ while the labeled transitions $rt_i \in RT$ are inside the ring.

4 CONCLUSION AND FUTURE WORK

Our results from Section 3 can be summarized in Table 1 and Table 2.

This paper quantified the number of nodes and arcs removed in the reachability graph after the application of Bride's reduction rules. Our technique focused on the markings and transitions affected by the removal of the workflow net nodes via each reduction rule. Knowing which reduction rule performs "better" may be used for fine tuning reduction rule based algorithms. The efficiency of other reachability rules can be analyzed using the methods used in this paper.

Table 1. Summary of order reduction where RG and $\overline{RG}$ are the reachability graph of $R(N, i^k)$ and $R(\overline{N}, i^k)$, respectively.

Reduction Rule	Order
$\phi_{RemoveP}$	$order(\overline{RG}) = order(RG)$
$\phi_{RemoveT}$	$order(\overline{RG}) = order(RG)$
$\phi_{RemoveST}$	$order(\overline{RG}) = order(RG)$
$\phi_{RemoveTP}$	$order(\overline{RG}) = order(RG) - |\{M : M \in R(N, i^k) \text{ and } M(p) > 0\}|$
$\phi_{RemovePT}$	$order(\overline{RG}) = order(RG) - |\{M : M \in R(N, i^k) \text{ and } M(p) > 0\}|$
$\phi_{RemoveR}$	$order(\overline{RG}) = order(RG)$
	$- |\{M : M \in R(N, i^k) \text{ and } \exists p' \in RP, M(p') > 0\}|$
	$+ |\{M : M \in \mu(R(N, i^k)) \text{ and } M(p) > 0\}|$

Table 2. Summary of size reduction where RG and $\overline{RG}$ are the reachability graph of $R(N, i^k)$ and $R(\overline{N}, i^k)$, respectively.

Reduction Rule	Size
$\phi_{RemoveP}$	$size(\overline{RG}) = size(RG)$
$\phi_{RemoveT}$	$size(\overline{RG}) = size(RG) - \lvert\{(M_i, M_j) : M_i \overset{t}{\longrightarrow} M_j\}\rvert$
$\phi_{RemoveST}$	$size(\overline{RG}) = size(RG) - \lvert\{M : M \in R(N, i^k) \text{ and } M \text{ enables } t\}\rvert$
$\phi_{RemoveTP}$	$size(\overline{RG}) = size(RG) - \lvert\{(M_x, M_y) : M_x \overset{t}{\longrightarrow} M_y\}\rvert$
	$\quad - \lvert\{(M_x, M_y) : \exists t' \in T\backslash t, (M_x(p) > 0 \text{ or } M_y(p) > 0)$
	$\quad \text{where } M_x \overset{t'}{\longrightarrow} M_y\}\rvert$
	$\quad + \lvert\{(M_x, M_y) : M_x(p) = M_y(p) = 0 \text{ and } \exists t' \in T\backslash\{t\},$
	$\quad \exists M', M_x \overset{t'}{\longrightarrow} M' \overset{t}{\longrightarrow} M_y\}\rvert$
$\phi_{RemovePT}$	$size(\overline{RG}) = size(RG) - \lvert\{(M_x, M_y) : M_x \overset{t}{\longrightarrow} M_y\}\rvert$
	$\quad - \lvert\{(M_x, M_y) : \exists t' \in T\backslash t, (M_x(p) > 0 \text{ or } M_y(p) > 0)$
	$\quad \text{where } M_x \overset{t'}{\longrightarrow} M_y\}\rvert$
	$\quad + \lvert\{(M_x, M_y) : M_x(p) = M_y(p) = 0 \text{ and } \exists it \in IT,$
	$\quad \exists M', M_x \overset{it}{\longrightarrow} M' \overset{t}{\longrightarrow} M_y\}\rvert$
$\phi_{RemoveR}$	$size(\overline{RG}) = size(RG) - \lvert\{(M_x, M_y) : M_x \overset{t}{\longrightarrow} M_y, t \in RT\}\rvert$

On the basis of these results, it may be useful to use combinatorial or algebraic methods to determine the exact amount of nodes and arcs reduced in the reachability graph. The equivalent algebraic operation of the reduction rules on the matrix representation of workflow nets will provide us an alternative approach in determining the efficiency of reduction methods.

REFERENCES

Bride, H., Kouchnarenko, O., and Peureux, F. (2017). Reduction of workflow nets for generalised soundess verification.

Desel, J. and Esparza, J. (1995). *Free Choice Petri Nets*. Cambridge University Press, Cambridge.

Lipton, R. (1976). The reachability problem requires exponential space. *Dep. CS, Yale Univ., Rep.*

Murata, T. (1989). Petri nets: Properties, analysis and applications. *Proceedings of the IEEE*, 77 (4):54–580.

Petri, C. A. (1962). *Kommunikation mit Automaten*. PhD thesis, Universität Hamburg.

van der Aalst, V. (1996). Structural characterization of sound workflow nets. *Computing Science Report Eindhoven University of Technology*.

Theory and Practice of Computation – Nishizaki et al (eds)
© *2021 Taylor & Francis Group, London, ISBN 978-0-367-41473-3*

Exploring a flexible blended learning model in technology deficient classroom

J.E. Gumalal, A.P. Vilbar & F.G. Bernardez
University of the Philippines Cebu, Cebu City, Philippines

ABSTRACT: The 21st-century learning integrates Information and Communication Technology (ICT) that aims to create a borderless education by extending learning outside the realms of a physical classroom. In implementing so, virtual learning environments (VLE) are often employed hand in hand with contemporary pedagogies such as the blended learning (BL) method. But the limitations brought by the realities of a technology deficient classroom in a developing country inhibits educators from entirely using full-on BL method in their curriculum. This exploratory action research study aims to determine the potential use of the Supplement, Engage, and Respond in a Virtual Environment (SERVE) blended model which was developed to address ICT integration in a technology deficient classroom. The model is also an addition to the vast array of blended learning methods, albeit being very customized to a specific skill and region. The experiential evaluation (through interviews and journaling) and physical origami outputs of 40 students from UP High School Cebu were collected and analyzed. SERVE was found to promotes ease, excitement and empowerment, and motivation in learning. However, SERVE must be delivered with high amount of teacher support. SERVE showed potential effectiveness through the quality of outputs the students created. However, a more quantifiable way of assessing the model's impact on students and satisfaction rating is needed to test its effectiveness and efficiency.

1 INTRODUCTION

The demands of the 21st century to develop multiliteracy and digital literacy among students have impacted education to use new but flexible methods in integrating information and communication technology (ICT)(Trilling & Fadel 2009). Adding to the reality is that learners are considered digital natives (McKnight 2018) who are enthusiastic about technological advances (Eastman & Liu 2012) and goal-oriented and motivated (Howe & Strauss 2003).

The teachers are required to supplement the learning process through computer-based instructional tools and to use blended learning (BL) methods to innovate traditional classrooms.

Using platforms such as a Learning Management System (LMS), educators around the world can create a virtual learning environment and extend their learning episodes anytime and anywhere (Koh et al. 2015). Many LMS can be accessed freely by teachers and students and are known to be evaluated highly on their usability, efficiency, effectiveness, and satisfactory interfaces (Kakasevski et al. 2008). This technology is best in a full-on BL method that global educators use. Theoretically, one can learn without borders. But this statement, however, may only be valid to some nations.

The physical and inherent educational limitations posed by socio-economic status of students in developing countries such as the Philippines may have caused educators, especially from public schools, to shy away from fully integrating ICT in their pedagogy (James 2005, Mardikyan et al. 2015). Public school students from low-income families do not have the technological and physical requirements, including the internet, at home to fully experience the borderlessness of their classrooms. These technical aspects must put into mind when integrating ICT or even using LMS (Bajaras & Owen 2000, Anderson 2008).

But all these limitations must not hinder a Filipino educator from innovating already established technological pedagogical models to suit the need of ICT integration. This study offers a flexible BL model that can be used in a technology deficient classroom as a response to this need. Hoping to add to the vast array of established BL models, the researchers' Supplement, engage and Respond in a Virtual Environment BL model aim to customize ICT use around the realities of a Philippine public school.

2 THEORETICAL FRAMEWORK

Blended learning (BL) is defined as a pedagogical method that mixes the practicality of the face-to-face (f2f) format and the capacities of an online learning platform to maximize individual learning and collaborative skills of a set of individuals (Carman 2002, Umoh & Akpan 2014, Hoic-Bozic et al. 2015, Akbarov et al. 2018). In this sense, it can be used to suit and promote education in a varied environment by using several methods. What allows BL models to do this? The model itself employs the eclectic use of learning theories under behaviorism, cognitivism, and constructivism.

Blended learning allows students to observe and imitate what they see and what they experience from their peers and teachers, which is a hallmark of a behaviorist approach (Wang 2011). It relies upon the learner's ability to gather, process, and reorganize information based on its cognitive structures or scaffolding, which is employing the cognitivist approach. Teachers, therefore, must acknowledge the need to divide the learning process into meaningful parts and organize them in logical complexity to aid the learner (Alonso et al. 2005). Not only a well-planned process essential in a BL method, but the use and availability of media and other sources to promote a more branched and broader view of learning, acknowledges that students can learn via problem-solving, discovery, and social interaction in a learning community or via social constructivism (Carman 2002, Hoic-Bozic et al. 2015).

3 LITERATURE REVIEW

Virtual Learning Environment (VLE) is a set of learning and teaching tools designed to incorporate computers and the internet in the learning process. It is often composed of a learning management system which contains curriculum maps, student activity tracking, electronic communications system, and online resources accessible anywhere by connecting to the internet and using either a Teacher ID or a Student ID.

Schools consider VLE as revolutionary and revitalizing (Selwyn 2007). In effect, many institutions created VLEs to aid instruction and to provide the students with an innovative way of learning without compromising their social interactions and private life (Lukman & Krainc 2012).

Teachers create their modules, link references, deliver and revise lessons, and even teach a large number of students within the VLE which means that the students would not necessarily follow a pre-determined set of experiences, and they have the ability to interact with their teachers and peers to do an evaluation or even feedback about their learning (Woessmann & West 2006). It means that the teacher will now have a chance to create various pedagogy based on the specific need of a subject or topic. In this manner, learning in f2f while integrating online learning is inevitable. In a way, the use of VLE also points into using a BL method.

With its characteristics in terms of space and collaboration, VLE is a great way to transition to more students and process-centered education (Lameras et al. 2012). BL in this style is also known to improve learning outcomes while allowing the teachers to lead a more customized learning scaffolding for their students through collaboration, discovery, and direct instruction. Some contemporary blended methods in an electronic course focused on personalized problem-based learning using Web 2.0 and E-learning activities Recommender systems that are used to automatically select varied activities for the students based on their current progress (Hoic-Bozic et al. 2015). Although this recommender or response system is excellent to use, most free LMS does not have this feature, and, evident later in this study's SERVE model; it is possible to make the activity recommendation manually. The idea may put-off educators who have the computing means of automating the customization of their student's learning, but this personal non-computer automated recommender system may become a reality for teachers in developing countries.

3.1 *ICT underutilization in developing countries and the problem it poses in VLEs*

There has always been a global divide in the ability of developing and developed countries in ICT driven services. Poorer countries have citizens with lesser ability to own computing and internet services, which may translate to computer illiteracy (Lameras et al. 2012). To address the problem, public schools develop computer laboratories that are shared by students, libraries offer free net searching, and internet shops offer pay-per-hour computer use. These are standard internet connectivity options for the general public. Students, however, must settle to share devices and teachers, who are teaching a minimum of forty students, must put up with the scheduling of projectors and, in some cases (like the researchers) opt to buy laptops to ease the scheduling mishaps personally. Although the presence and availability of these technological resources may contradict the notion that developing countries do not have technological innovations, it does not take away that these countries have tech-driven diffidence (James 2005). But this constraint does not hinder institutions and teachers from trying to use LMSs and creating VLE.

In a study in Brazil, for example, tried to institutionalize the use of VLE using a BL method but found that the technique is entirely new to their community (Ribeiro et al. 2017). They also reported that aside from the availability of equipment, the enrolment of the students to the VLE is also dependent upon the teacher's ability to use the VLE. They also reported that the potentials of VLE use are still predominantly high if there is an avenue for inclusion. In this regard, a teacher who adheres to institutionalizing the use of VLE must be ready to encourage their students to enroll and try the new method.

A study by Clarida and team (2016) shows that organizational factors, elements of the course, VLE's navigational capacities, and other intrinsic factors such as student/teachers' technological skills and ability to critically solve problems regarding their use of the platform may drive digital exclusion. In addition to that, a study in the University of Uyo in Nigeria pointed out that the non-availability and non-accessibility along with poor ICT skills in ICT of students hamper the utilization of e-learning in teaching and learning (Umoh & Akpan 2014). ICT skills and infrastructure must not only be attained by the students but also by the management and educators of developing countries (Kituyi & Tusubira 2013). However, good VLE project borderless education to address equity in learning, limitations on the infrastructure, hard wares, and connectivity often plague the platform (Mardikyan et al. 2015). These reports may look unappealing to a teacher from a developing country, but it only shows that it is even more important to try and continue the advocacy for digital literacy.

The utilization of the BL method may answer the limitations mentioned above. BL method for graphic design for developing countries of Atef & Medhat (2015), for example, utilizes 40 percent f2f, 20 percent social interactions, 20 percent virtual class, and 20 percent self-study emphasizes that the key to incorporating technology is the internal expertise of educators who can revise the curriculum based on the available infrastructure and connections. In addition to that, Mtebe & Raphael (2013) stressed that institutions must be wary of outdated learning resources, underutilized learning centers, and technical difficulties that may appear in a BL method. What do all this mean? Developing countries must have institutions and educators who are willing to build and continue to revise pedagogical methods to implement ICT integration in their respective regions eventually. Not only will this impact the national system, but it will also increase the computing capacity of the students as they transition towards the global workforce. Thus, the creation of a flexible blended learning model is needed.

4 OBJECTIVES OF THE STUDY

This research aims to develop a blended learning model that addresses the needs of the students who have limited internet connection at home. Referred to as a Supplement, Engage, and Respond in a Virtual Environment, the model aims to provide an engaging physical or virtual classroom environment that supplements the learning needs of the students through online resources with feedbacking with the teaches and peers. Furthermore, this research aims to determine the student's opinions and experiential evaluation of the model.

5 RESEARCH METHODOLOGY

This research used the qualitative method of developing and evaluating the SERVE model. It analyzed studies and literature on blended learning to develop the model and used reflective journals and focus group discussion on evaluating the model.

5.1 *About the study*

The study involved the whole population of Grade 8 students of the University of the Philippines High School Cebu because they exhibit the characteristics needed to determine the applicability of the model. They come from low-income families, and they rely upon the school's computer laboratory and nearby internet cafés for internet connectivity.

Although most of the students in this High School have smartphones, many of them have limited access to the internet at home due to economic reasons. In this study, only five students have internet access at home.

5.2 *Application of the SERVE model to the arts class*

The students used the UP Cebu Virtual Learning Environment (UP Cebu VLE), which is a Moodle-based platform (an LMS). The SERVE Model was implemented in the Grade 8 Asian Arts study which focused on Origami for one grading period. The students learned origami for one grading period utilizing this platform with specified facilitation by the teacher.

First, the teacher created modules with specific lessons to supplement the self-study or practice of the students. These modules were uploaded online and embedded in the self-paced sections inside the UP Cebu VLE to scaffold the learning of the students. The activities ranged from simple to complex. The teacher then enrolled all of the students in the VLE, specifically in the Asian Arts and Basic Painting course site. Figure 2 shows the procedure in creating the VLE account, studying the module, and receiving the feedback.

The class instruction inside the computer laboratory only happens for an hour. The students learn the origami folding by imitating the designs shown in the selected online videos and step-by-step manuals. Maintaining a useful and meaningful online material is key to effective e-learning (Kituyi & Tusubira 2013). The method adheres to the cognitivist approach of scaffolding the learning episodes and the behaviorist approach's learning by imitation (see Figure 1).

Second, the students engaged with their peers through a forum and to their teacher through f2f evaluation. The students asked questions online or personally when needed. They answered assessment tools and were encouraged to share their answer and teach other students. Following the constructivist approach, the students learn by interacting with peers and are trained to receive and give feedbacks which extends their learning episodes through inquiry and discovery (Carma 2002).

Lastly, the teacher responded to the learning progress of the students. Because students learn a physical skill, the teacher properly schedule the delivery of the module and the collection of tangible outputs while taking into mind the feedback of the students about the lessons (Poon 2012). The analysis of the origami outputs of the students prompts the teacher to revise and review the planned activities and customize the module according to the students' progress, a responsive method of supplementation like that of a recommender system.

Figure 1. Students use the videos and modules in the UP Cebu VLE to learn the origami design.

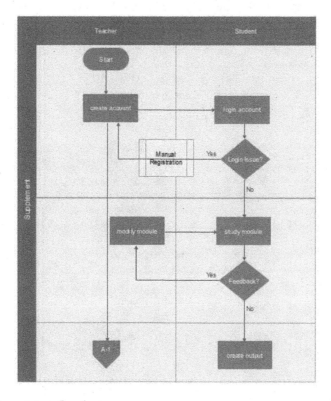

Figure 2. Supplement stage flowchart.

The students passed their origami outputs for evaluation, which allow teachers and students to do away with digital deficiency in accessing the modules (Woessmann & West 2006, Atef & Medhat 2015). The idea comes from the fact that students still prefer to take paper and pencil test or pass assignments personally (Akbarov et al. 2018). Requiring the students to do digital work at home is not encouraged due to infrastructure and equipment limitation. Instead, the scheduled e-learning and traditional learning in school are optimized. Atef & Medhat first proposed this idea of a flexible blended learning model in 2015.

Investigation on the folding behavior of the students showed that they were able to grasp the primary competency of the topic, which is to understand how the folds behave and create an original origami model.

The students who varying levels of creative design based on the Taxonomy of Creative Design by Nilsson (2011). These levels can be seen in Figure 3. Acquiring the highest level in the taxonomy means that the students were able to grasp the needed skills in making and designing their origami which would mean that the used SERVE blended learning model is a possible method in teaching origami.

6   EVALUATION OF THE SERVE MODEL

After using the SERVE Model for one grading period, the students claimed that the model developed faster learning, excitement and novelty, motivation to learn, and cost efficiency. These claims were taken from the reflective journals and focus group discussions.

6.1   *SERVE promotes ease in learning*

The SERVE model eases the learning process making students learn quickly. According to Archie, learning through the computer is faster and neater. He added, "Following instructions that are meant to be read and with pictures supplied is less stressful than trying to catch up with a discussion. I don't dislike anything about this concept!". Bernard also said, "I learn fast by following the videos."

The statements of the students prove that blended learning model with the availability of a VLE as its primary learning platform is enough if the students can quickly feel its usability,

Figure 3.   Students creation under imitation (top), transformation (middle) and origination (bottom).

mechanism of engagement, and ease of use (Umoh & Akpan 2014, Clarida et al. 2016, Akbarov et al. 2018). The model must make the student feel the freedom and openness of an online learning method.

6.2 *SERVE promotes learning for the new generation*

Archie, aside from the fact that he liked the VLE's ease of use, was very excited and it is easy to see why. Students like Archie and Bernard are born for this new type of learning. The traditional factors for the digital exclusion (e.g., Sex and age among others) seem to be less apparent in this situation as the students, being digital natives, have an inherent belief of their capacity to learn and use a specific technology (Clarida at al. 2016) which was evident in Cassie's journal entry saying "Well, having a new method in teaching makes it easier, because our generation is in technology already. Where people would easily learn new things through the internet." This statement was indeed one of the many reasons why e-learning was encouraged in 21st-century education. Students are willing to try and to learn electronically (Rotherham & Willingham 2010).

Aside from Cassie, Daniela also curiously stated, "I feel that it is easy for me to use the computer even if we don't have any at home."

In addition to their statements, Emily also said, "I felt overwhelmed because learning through technology is not everywhere, some schools don't [sic] have any computer laboratories due to financial problems, and if you don't have any computer at home, at least you can learn computer at school." The statement gives off the feeling that the model was not only suited for the students but also very empowering. Since the students come from families with lower socioeconomic status, they often have insecurities when using technology in learning, but it doesn't mean that they are unwilling to try. The behavior was also found in the studies of James (2005) and Tarus et al. (2015).

6.3 *SERVE promotes learning motivation*

Students also reported that they felt more motivated to learn and ecstatic, especially when browsing other sources and being supported on and off the VLE platform. Open peer collaboration encourages the creation of a supportive community within the student's online environment. This set up help students who are studying alone to be more motivated to finish their class works (Woessmann & West 2006, Biddle & Berliner 2008, Lukman & Krainc 2012, Kituyi & Tusubira 2013).

Fatima, for example, said that to be able to see the work of other students and to take credit for her work makes her feel motivated. Although some, unlike Fatima, would want only their teacher to see their work. Hannah, in her reflective journal, said she feels "makuyawan (fearful)" and "mauwaw (shy)" when people see her works. She also stated that "technology is good, but it may cause some students to be lazy."

6.4 *SERVE must be used with a high amount of support*

Some felt that being taught and demonstrated by their teacher entirely was better than watching videos and just following. Graham reiterated, "I like more when the teacher demonstrates to us. I want someone to teach me."

This statement only emphasizes that teachers utilizing any blended learning methods must check on the students' progress personally and offer support (Selwyn 2007). The teacher's presence in both online and offline settings is one of the factors of student e-learning satisfaction (Tarus et al. 2015, Nortvig et al. 2018). Teachers who wish to use the SERVE model must be present while the students learn.

Overall, students feel excited and fulfilled in terms of technological improvement. They also understand that the SERVE model was exploratory and may result in some students, either becoming lazy or become more efficient. Use of online learning mechanism entails that the platform and its contents must be developed and revised regularly to accommodate changing behaviors of students in e-learning (Tarus et al. 2015, Hoic-Bozic et al. 2015, Atef & Medhat 2015, Nortvig et al. 2018). But is SERVE a perfect blended learning model? The answer is no, but its positive reception means that it can still be developed further. These and other possibilities were hoped to be addressed as the model undergoes further evaluation.

7 CONCLUSION AND RECOMMENDATION

Teaching practice-driven skill such as origami may use flexible blended learning in technology deficient learning environments by navigating around the current set of limitations while applying pedagogical theories available for such situation (in an LMS/VLE).

The students viewed the SERVE blended learning model as an innovative way of learning. The study also acknowledged that most students do not have access to computers and the internet at home, and some students needed direct instruction in learning. Since the model was exploratory, it may result in some unique situations or learning habits.

The students were able to learn while collaborating or connecting with fellow learners and learning individually through direct instruction. However, teacher presence was highly advised during their learning episodes.

The student's outputs verified their understanding and development of origami skills using the SERVE model. It is recommended that a more quantifiable way of assessing the model's impact on students and satisfaction rating is needed to test its effectiveness and efficiency.

REFERENCES

Akbarov, A., Gonen, K., & Aydogan, H. 2018. Students' Attitudes Toward Blended Learning in EFL Context. *Acta Didactia Napocensia*: 11(1), 61–68.

Alonso, F., Lopez, G., Manrique, D., & Viñes, J. M. 2005. An instructional model for web-based e-learning education with a blended learning process approach. *British Journal of Educational Technology*: 36 (2), 217–235.

Anderson, T. 2008. Social software to support distance education learners. *The theory and practice of online learning*, 221.

Atef, H., & Medhat, M. 2015. Blended Learning Possibilities in Enhancing Education, Training, and Development in Developing Countries: A case study in Graphic Design Courses. *TEM Journal*: 4(4), 358–365.

Bajaras, M., & Owen, M. 2000. Implementing virtual learning environments: Looking for a holistic approach. *Educational Technology & Society*: 3(3), 39–53.

Biddle, B. J., & Berliner, D. C. 2008. Small class size and its effects. *Schools and Society: A Sociological Approach to Education*: 3, 86–95.

Carman, J. M. 2002. *Blended learning design: Five key ingredients.*

Clarida, B. H., Bobeva, M., Hutchings, M., & Taylor, J. 2016. Strategies for Digital Inclusion: Towards a Pedagogy for Embracing and Sustaining Student Diversity and Engagement with Online Learning. *IAFOR Journal of Education*: 86–106.

Eastman, J. K., & Liu, J. 2012. The impact of generational cohorts on status consumption: an exploratory look at generational cohort and demographics on status consumption, *Journal of Consumer Marketing*: 29(2): 93–102.

Hoic-Bozic, N., Dlab, M. H., & Mornar, V. 2015. Recommender System and Web 2.0 Tools to Enhance a Blended Learning Model. *IEEE Transactions in Education*. doi:0.1109/TE.2015.2427116

James, J. 2005. The global digital divide on the Internet: developed countries constructs and Third World realities. *Journal of Information Science*: 31(2), 114–123.

Kakasevski, G., Mihajlov, M., Arsenovski, S., & Chungurski, S. 2008, June. Evaluating usability in learning management system moodle. *ITI*: 30.

Kituyi, G., & Tusubira, I. 2013. A framework for the Integration of e-learning in Higher Education Institutions in Developing Countries. *International Journal of Education and Development using Information and Communication Technology*: 9(2), 19–36.

Koh, J. L., Chai, C. S., Benjamin, W., & Hong, H. Y. 2015. Technological Pedagogical Content Knowledge (TPACK) and design thinking: A framework to support ICT lesson design for 21st-century learning. *The Asia-Pacific Education Researcher*: 24(3), 535–543.

Lameras, P., Levy, P., Paraskakis, I., & Webber, S. 2012. Blended university teaching using virtual learning environments: conceptions and approaches. *Instructional Science*: 40(1), 141–157.

Land, M. H. 2013. Full STEAM ahead: The benefits of-of integrating arts into STEM. *Procedia Computer Science*: 20, 547–552.

Lukman, R., & Krainc, M. 2012. Exploring Non-Traditional Learning Methods in Virtual Methods in Virtual and Real World Environments. *Journal of Educational Technology & Society*: 15(1).

Machado, M., & Tao, E. 2007. Blackboard vs. Moodle: Comparing user experience of learning management systems. *Frontiers in education conference-global engineering: knowledge without borders, opportunities without passports, 2007. FIE'07. 37th annual* (pp. S4J–7). IEEE.

Mardikyan, S., Yildiz, E. A., Ordu, M. D., & Simsek, B. 2015. Examining the global digital divide: a cross-country analysis. *Communications of the IBIMA, 2015*: 1.

McKnight, K. 2018. The first generation of true digital natives. *Research World*: 2018(70), 14–17. doi: 10.1002/rwm3.20659

Mtebe, J., & Raphael, C. 2013. Students' Experiences and Challenges of Blended Learning at the University of Dar es Salaam, Tanzania. *International Journal of Education and Development using Information and Communication Technology*: 9(3), 124–136.

Nilsson, P. 2011. The challenge of innovation. In critical thinking and creativity: learning outside the box. *9th International Conference of Bilkent University Graduate School of Education*:54-62). Ankara, Turkey: Bilkent University.

Nortvig, A.-M., Petersen, A., & Balle, S. H. 2018. A Literature Review of the Factors Influencing E-Learning and Blended Learning in Relation to Learning Outcomes, Student satisfaction, and Engagement. *The Electronic Journal of e-learning*: 16(1), 46–55.

Parsad, B., & Spiegelman, M. 2012. Arts Education in Public Elementary and Secondary Schools: 1999-2000 and 2009-10. *National Center for Education and Statistics. NCES* 2012-14.

Poon, J. 2012. Use of blended learning to enhance the student learning experience and engagement in property education. *Property Management*: 30(2), 129–156.

Ribeiro, A. d., Oliveira, E. R., & Mello, R. F. 2017, January-March. Building a Virtual Learning Environment to foster blended learning experiences in an Institute of Application in Brazil. *Open Praxis*: 9(1), 109–120.

Rotherham, A. J., & Willingham, D. T. 2010. 21st-Century Skills. *American Educator*: 17, 17–20.

Selwyn, N., 2007. The use of computer technology in university teaching and learning: a critical perspective. *Journal of computer assisted learning*: 23(2), 83–94.

Tarus, J., Gichoya, D., & Muumbo, A. 2015. Challenges of Implementing E-learning in Kenya: A case of Kenyan Public Universities. *International Review of Research in Open and Distributed Learning*: 16(1), 120–141.

Trilling, B., & Fadel, C. 2009. 21st-century skills: Learning for life in our times. *John Wiley & Sons*.

Umoh, J., & Akpan, E. 2014. Challenges of Blended E-Learning Tools in Mathematics: Students' Perspectives University of Uyo. *Journal of Education and Learning*: 3(4), 60–70.

Wang, Y. 2011. Research and instructional practice based on blended learning. *Proquest Dissertations and Theses Global*.Retrieved from https://search.proquest.com/docview/1870671060?accountid=47253

Wares, A., 2013. Appreciation of mathematics through origami. *International Journal of Mathematical Education in Science and Technology*: 44(2), 227–283.

Woessmann, L., & West, M. 2006. Class-size effects in school systems around the world: Evidence from between-grade variations in TIMMS. *European Economic Review*: 50(3), 695–736.

Theory and Practice of Computation – Nishizaki et al (eds)
© *2021 Taylor & Francis Group, London, ISBN 978-0-367-41473-3*

Developing Dark Night 2: Analyzing the experience of blind users with a mobile role-playing game

Hans Gustaf G. Capiral, Kenneth V. Velasquez, Ma. R. Solamo & R.P. Feria
University of the Philippines Diliman, Philippines

ABSTRACT: This research investigated how the visually-impaired individuals interact with mobile Role-Playing Games (RPGs). Dark Night 2 was developed using only auditory and haptic components, and was tested by visually-impaired individuals. They were observed during gameplay, and were interviewed. Their feedback and suggestions were analyzed. The results, aside from showing that they also enjoy playing RPGs, reveal four major themes to improvement the interaction of visually-impaired individuals in playing RPGs to further enhance their experience. They are *using high quality audio as a medium*, *having a deeper story-line*, *having a right amount of game challenge*, and *using comfortable and intuitive controls*.

1 INTRODUCTION

In our modern age, people use their smartphones often. Some use it for official businesses. Others use it for entertainment, such as music and games. Advancements in technology shaped our present into something that is mostly experienced visually, which became a challenge to those who are visually-impaired. Fortunately, technology is also advanced enough to enable them to use the same technology (**Csapo2015**) albeit in a limited manner. Thus, the visually-impaired are not experiencing the full capability of today's technology, especially in games, which is oftentimes a mix of visual and auditory experiences. They also lack adequate training to allow them to utilize the potential of these technologies (**gerber2003-benefits**) Inspired by the game developed by Dowino, called "A Blind Legend", the researchers have thought about the possibility of turning games from a visual-auditory experience to a primarily auditory one with a minor haptic component, particularly for the genre of role-playing games (RPGs). RGPs are mostly enjoyed due to the visual satisfaction they provide. This research investigates how the visually-impaired individuals experience playing an RPG running on a mobile devices using only audio and haptic components.

2 DARK NIGHT 2

Dark Night 2 is an adventure role-playing game (RPG) developed mainly using audio and haptic as the main avenue for game interaction on a mobile device. In the game, the user plays as a blind mercenary in prison. The mercenary will be helped by an unknown individual and will be tasked to help common people while staying away from guards, lest the mercenary be captured and put back in prison. Along the way, the mercenary will know the reason why he was put in prison and how he can clear his name and reunite with his group.

The player must navigate through the levels, following audio cues by touching and dragging on the screen to move the player. If the player encounters an enemy, the player will be engaged in a battle mode, where the player must swipe on the screen to damage the enemy he encountered. By swiping at the right time, the player will be able to deal a critical damage and prevent the enemy from inflicting damage on him.

Dark Night 2 is divided into two parts. The first part focuses on the game teaching the player how to play the game. This mainly includes the helper telling the player the rules and what to do during the level. This serves as the tutorial part of the game. The next part lets the player explore the game by doing quests. But this time, there is no the guidance from the helper. The player only gets audio and haptic cues from the game objects. The general layout of the game world can be seen in Figure 1. The boxes represent the levels. They were designed to be adjacent to eliminate turns, thus, making navigation easier.

Dark Night 2 was developed using Unity, a cross-platform game engine. It utilizes the 3D sound engine, which have the directional sounds allowing easy implementation, and requiring little modifications to the sound in order fit the purpose of the game. It also utilizes the built-in handheld vibrations for the necessary tactile sensation coming from the game.

Dark Night 2 was tested using Lenovo A5000 running Android 4.4.2 (Kitkat). The device has a 3.5 mm headphone jack, allowing for usage of stereo-enabled headphones/earphones, which will be integral for the directional location of audio sources. It is also equipped with 5-point multi-touch and haptic feedback.

3 ASSUMPTIONS INFLUENCING GAME DESIGN AND DEVELOPMENT

Many assumptions were made that had a big impact on the direction of the development and testing of Dark Night 2. Since the target users are visually-impaired individuals, a major assumption is to forgo the visual design and development of the game. Rather, the design and development of the game focused more on audio and haptic components.

3.1 *Assumptions on sound components*

The use of in-game tutorials are a great way of teaching players how to play the game while sticking to the storyline. Guidelines on developing games with the focus on visually impaired (**jaramillo2017mobile**) suggests that in-game tutorials in the form of narration is necessary to introduce the player on the narrative of the game. Thus, a narration was provided at the start and tutorial of the game.

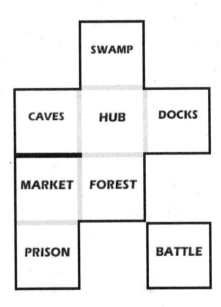

Figure 1.   Game World Design of Dark Night.

Background noises or ambient sounds enhance the atmosphere of gameplay by helping the players immerse themselves in the game (**archambault2007computer**) Several sound assets were used to fit the theme of the level the player is on. They are significantly different from each other so that the players can easily distinguish that they are on a different level. For example, the ambience in the first level gives an atmosphere of a medieval prison while the next level gives an ambience of being in a market.

Due to the assumption of the visually-impaired users having difficulty navigating through unfamiliar environments (**mothiravally2014attitude**) guide sounds were used to help players navigate through the levels. They are bound to a game object that helps guide the player to the target location. They use repetitive sounds that are distinguishable from the ambient sounds of the environment. They must be repetitive and easily distinguishable from the surrounding sounds. This was achieved through the use of volume control and tonal sounds rather than a musical one (**mereu1997audio**) For example, the player must navigate through the level by following the sound of bells. The closer the player is to the bell, the louder it rings. A good balance between the guide sound and the background audio allows the player to distinctively hear the guide sound over a long distance.

Usage of earphones in playing the game to determine the directional location of game objects was also one of the key assumptions. Using earphones allowed the players to easily determine the direction where certain sounds were coming from. It was assumed that sounds from the four main directions, i.e., left, right, front and back, were be easily distinguishable. However, from the four directions, only left and right were easily distinguishable. Both the front and back were classified as coming from the middle.

3.2 *Assumptions on haptic components*

The research on how to use tactile channel to give information on the layout of the screen to the blind on a smartphone **buzzi2013haptic** brought about the idea that haptics can also convey information, and at times, reinforce the importance of the information conveyed through other channels. However, using haptics provided challenges in designing the application, particularly, where and when it is appropriate to trigger a vibration. It also presented problems in testing, since the researchers cannot easily observe the state of the haptic feedback, particularly, on whether the player is having problems.

Gesture patterns used to interact with the screen of the mobile device affects the user experience of the player. Simple gestures like one finger swipes and taps are more preferable than those that use letter or shape-based patterns (**buzzi2015exploring**) Following this, the game controls were designed as simple as possible to prevent frustration and to deliver smoother gameplay. The use of a virtual joystick in Figure 2 simplifies navigation controls for the player.

Cues, whether through sound or vibration, were also included under the assumption that they may serve to give feedback to the player. Audio cues, and sometimes haptic cues, are bound to game objects to give the player an idea that he is interacting with a certain game object. An example of such interaction is the player colliding into a wall in the game world. It

Figure 2.   Virtual Joystick.

triggers a sound indicating a wall was hit, and at the same time, triggers the phone to vibrate as well.

4   TEST METHODOLOGY

Due to the limited population of the target test participants, a qualitative approach in investigating the interaction of visually-impaired individuals using RPG on mobile phones was performed. It focused on understanding their experiences on playing RPGs on mobile devices. Testing was done in a controlled environment with minimal researcher intervention. Data gathering methods used were *observation* where players are observed during game play, and *interview* where they answer a few questions regarding the application and their experience in playing it. Body language, feedbacks, and interview results were the sources of information.

The testers were observed using the Colin Robson Framework (**robson2016real**) This framework allows for a detailed observation that prevents overlooking information relevant to the topic at hand. It has the following terms that aid the researchers in remembering how to make notes. Namely, they are space, actors, activities, objects, acts, events, goals, and feelings. *Space* involves the place where the participants are interacting in. *Actors* involves the participants themselves. *Objects* involves the physical objects present relevant to the observation. *Activities* involve what the actors are doing. *Acts* pertain to what specific individuals are doing. *Events* pertain to the events that may have induced the certain acts observed. *Goals* pertain to what the actors are trying to accomplish. *Feelings* pertain to the mood of the group and/or individuals.

4.1   Test participants

The test participants were limited to only three (3) due to the limited nature of the population of the targeted visually-impaired people who were willing to take part in the research. Most of the participants came from Adaptive Technology for Rehabilitation, Integration, and Empowerment of the Visually Impaired (ATRIEV) learning center. It is a school that teaches the visually-impaired how to interact with computers. The other was recommended by the Persons with Disability (PWD) Sector of the local government of Antipolo City.

4.2   Test activities

4.2.1   Activities before testing
The first testing was conducted in the house of the participant recommended the PWD Section of the local government of Antipolo, City. For the environment, three (3) chairs were set up where the tester and the researchers were seated. The contents of the consent form were read and explained to the tester. He was, then, asked to sign if he agrees to the contents of the consent form. Testing commenced after he signed.

The second testing was conducted at ATRIEV administration office, where the researchers were asked to sign in a logbook for conducting tests with their students. They were also asked to read the guidelines in interacting with the disabled. They, then, proceeded to the classroom alongside the teacher of the class where the testing will be taking place. They were introduced by the teacher and the teacher asked his students who were willing to participate in the test. In the classroom, a table with two chairs were set up where the tester and one of the researchers were seated. A charging outlet was placed for the testing device. Much like the first testing, the contents of the consent form was read and explained. Tester were asked if they were willing to participate, they needed to sign the said form.

4.2.2   Activities during testing
Testing Dark Night 2 involved the testers playing the game. It included most of the game's mechanics under the introduction to the story only. This was done to address the issue of

testing taking up a lot of the testers' time. The researchers actively observed the testers, taking into account Colin Robson Framework while at the same time, video recording was on-going.

4.2.3 *Activities after testing*

The interview part of the test required the testers to answer a few questions regarding the game that they tested. The questions address the appropriateness of the features that the game *contained* and the game itself. Feedbacks regarding the game and how to improve it are encouraged to elicit reactions and to provide further information on the tester's perspective and experience on playing an RPG game.

4.3 *Test instrument*

To extract data for the testing, the researchers have settled on videotaping both the testing and the interview coupled with note-taking. These video recordings were reviewed multiple times to come up with more accurate and consistent findings in order to draw up conclusions regarding the experiences of visually-impaired when playing RPGs. Specific body language cues were observed to discern various information regarding the testers and their perspective on the game.

Interview Questions allowed getting information on specific aspects of the role-playing game. Data regarding the use of audio, haptics, and the game proper were gathered to see whether the design decisions were appropriate and which ones needed improvements.

4.4 *Data analysis*

The data gathered from the observation and interview were analyzed using thematic analysis on qualitative data. Patterns will be drawn on the behavior of the participants while they were playing the game. Parts of the game where the participants had taken a long time to finish as well as their body language during this time were essential for this type of analysis.

5 RESEARCH OBSERVATIONS

The participants of the test consist of three (3) visually-impaired individuals. Two of them have experienced audio-centric mobile games while the other has not. Audio-centric mobile games, usually in the form of interactable stories, have a linear story line. Dark Night 2 is an audio-centric mobile game with a nonlinear story line, meaning that the player controls how the story progresses.

The first tester that finished the test was a totally blind individual contacted through the help of the PWD sector of local government of Antipolo City. *Dominic* (not his real name) does not have any experience playing video games on a smartphone, but has experience using one for communications. Due to his lack of experience of playing video games on a mobile device, the context of the information gathered from him was different than the other two participants. Aside from experiences in playing mobile games, he is also older than the other two, bringing a different perspective on the study.

The next two participants were two totally blind students from the Adaptive Technology for Rehabilitation, Integration, and Empowerment of the Visually Impaired (ATRIEV) learning center. Philip and Ian, (not their real names) already had some experience with audio-centric mobile games, and that they were actually taking an Introduction to Programming class. These details were mentioned with great emphasis, as our findings regarding their experiences and feedback would most likely be attributed to these characteristics.

5.1   Experience for tester examiner Dominic

During the test, Dominic can be seen to have familiarity in his environment. Getting to his chair, the researchers observed him touching furnitures and walls to navigate towards his place. His posture shows relaxation once he got to the chair.

During the playthrough, one of the things immediately noticeable was the way that he controlled the character in the game. By observing the way that he moved around, getting to his chair, the researchers noticed the way how he applied real life movements to the game. Having little idea about the map, he stuck to the walls of the level, using it to give him an idea of where to navigate by letting the walls guide him. He was a bit perplexed by the audio of the wall bumps, but soon got used to it and used it to give him an idea of the presence of a wall and that he could navigate alongside it.

Later on, he was seen with changes in posture, from a leaning backwards to leaning forwards in his chair. Frustration was hinted at his expression, explaining that he hard a time navigating through the level because of the noise distractions. Also, he faced an issue on the controls of the phone. The problem was that one of his fingers holding the smartphone was touching the screen, resulting in his intended movements not being captured by the game (see Figure 3). He eventually readjusts his grip and fixed the problem.

In the interview, when asked if the sounds gave him an idea of the proximity of the objects, he agreed that sounds are a good way to indicate proximity of game objects. He made comments about how the audio helps him navigate through the level, especially the audio cues about the presence of walls. When asked about annoying sounds in game, he responded that there are none at all. When asked about the vibrations in game, he said that the vibrations are good in delivering information like wall collision and enemy attack prompt, and that variation in these helped distinguish different information for the player. Regarding the game, he said it was challenging, probably due to the fact that he had a bit of a difficulty following the guide in another level, but it was fun. He also said that the controls were fine, and the instructions were good since it eliminated the need for outside help in playing the game. At the end of the interview, he asked about the purpose of the game.

5.2   Experience for tester examiner Philip

During Philip's playthrough, one very obvious observation was his immense focus on the game. While playing, he occasionally fixed his posture, sometimes leaning back on his chair. Halfway through the test, he also became a bit more talkative even during times when he had difficulty navigating through the map. He talked about how he cannot progress through the second level, which can be interpreted as a way to distract himself from the frustration of being unable to move forward.

Philip commonly made a gesture while navigating, which was rapidly and repeatedly swiping the screen towards the direction he wanted to go. While the researchers did not design the navigation to work in a way that the in-game character moves faster depending on the speed

Figure 3.   False Touch Example.

of swipe gestures, this may be seen as an act of experimentation wherein Philip was looking for a way to increase the speed of the in-game character.

While there is no proper baseline to compare the speed of Philip's playthrough with, he took a significant amount of time (a little more than 10 minutes) navigating through the market level wherein his objective was to follow a repeating bell sound which guides him to the next level. This part of the test raised a very important issue which most likely caused his difficulty in navigation: the automatic rotation of the phone's screen. Due to not holding the phone upright at all times, the screen orientation at times swapped from portrait mode to landscape and vice versa. Because Philip cannot see the screen, he would not have been able to tell when the screen orientation swaps, let alone if it even swaps at all. This is a mistake hugely in part of the researchers, as they have overlooked this detail during the development of the application and there was no warning or reminder within the game itself to keep the device upright while playing, or to lock the orientation.

Unfortunately, Philip decided to quit testing before the battle level due to him being unable to move through the second level. His feedback on both during the playthrough and the post-test interview on this part of the game would not include.

From his interview, Philip mentioned several points and suggestions while answering our questions. When asked about the helpfulness of the sounds to deliver information about proximity, he mentioned that the in-game character's step sounds should match the environment in the game, for example the steps should sound like walking on pavement when navigating through the city. He also mentioned that the indicator sound for collisions should be changed because it sounds like he is giving a wrong answer to a quiz whenever he hits a wall. He suggested that it should be changed to a more realistic sound, where he can actually hear the in-game character bump into a wall. With regards to vibration, he said that it was effective but adding a menu where he could turn it off would be useful. When asked about the difficulty of the game, he said that it was mainly difficult due to the bell sound being too soft (or low) in volume, as well as the automatic screen rotation issue mentioned earlier. Finally, Philip said that he would not need additional help from others when playing the game.

5.3   Experience of tester examiner Ian

Compared to Philip's, Ian's playthrough went a lot smoother and faster, taking only about 5 minutes in total. This may be due to the nature of the environment, which was a single classroom, meaning that by witnessing the previous tester, Ian has already been made aware of the previous screen rotation issue that Philip faced in his playthrough, as well as the main objectives to accomplish in the game. Ian also appeared to be a bit more relaxed, as he kept the same posture from start to finish.

An issue that Ian faced was the lack of responsiveness in one part of the test, where he was swiping the screen but the in-game character would not move. However, this later on was classified by the researchers to be a hardware problem, as the device would sometimes not recognize touch input when the screen is not pressed hard enough. As such, it should not be reflected in the negative experience or feedback it brought with regards to the game itself. Nonetheless, this issue was again mentioned later in the interview, with Ian saying that in situations where the player would have to react quickly, this problem would serve as a great inconvenience. Aside from this, Ian did not seem to stumble upon any other issues as he finished the test.

In Ian's interview, he also mentioned Philip's suggestion about the step sounds, that it should match the current environment. However, unlike Philip, when asked about the helpfulness of the sounds in delivering information about proximity, he said that they were indeed useful and were able to give him a general sense of direction of the game objects. A few exceptions regarding the usefulness of sounds are the wall collision sound and the heartbeat sound during battle. For the wall collision, Ian said that he never really bumped into any walls, so he could not give any feedback regarding it, and for the heartbeats, he said that it was not very clear because even if he was able to hear the heartbeats getting faster as he loses health, there was no clear indicator sound for when his character was getting struck by the enemy.

About the background sounds for ambience, he said that they were very useful since they can help the player better imagine the place that the in-game character is in, especially if given the right *mix*.

For the vibrations, he specifically suggested that in the battle level, the device should vibrate when the enemy initiates an attack, which would notify the player so that they may be able to get ready for it. He also mentioned that different types of vibrations are definitely useful in conveying different types of information.

When asked about his frustrations during the test, he said there was only one, where he suddenly"died" without being hit. In reality, the game was actually programmed to restart upon clearing the battle level, so perhaps Ian misunderstood this by thinking that his in-game character was killed by the enemy. Again, this observation further supports the need for more and clearer indicators for various information.

When asked if he would need additional help from others when playing the game, he said that he did not need help because he already had some experience with games such as this. As for recommending the game to others, he said that he would do it once the problems were fixed, and also once the game had been given a grand purpose for the in-game character, presenting a more intriguing and interesting storyline compared to the bare one it currently has. He also mentioned that the game should not feel like "mini-games" but instead feel like a "real RPG". Finally, his last suggestion was to use better quality sound effects.

6   RESEARCH FINDINGS

Using thematic analysis on the consolidated results of the observations and interview of the three test participants, four (4) most prominent themes emerged, specifically, *using of high quality audio as a medium, having a deeper storyline, having the right amount of game challenge* and *Using of comfortable and intuitive controls.*

6.1   *Using of high quality audio as a medium*

Looking at the answers from the three testers, it can be seen that choosing audio as a medium for delivering game information is feasible provided that the quality of the sound assets should be realistic and should have the right mixture of sounds to create a vivid ambience of the place. By substituting sound assets for visuals, it is possible to create an engaging RPG that the visually-impaired will enjoy. Following this logic, high quality sound assets are akin to having high quality visuals in an RPG. They result in deep immersion for the players, which can be observed on how focus they are while playing the Dark Night 2.

Comparing the three testers' feedback regarding how effective a medium audio was, only Philip answered that the way it was implemented in the game was not very good, saying, that the sound assets needed to be a lot more realistic. Ian, on the other hand, said that it was effective, but also making the remark that the sound quality needs to be better as well. From what can be drawn from these answers, it can said that using audio as a medium for delivering game information would be feasible given a high audio quality for the sound assets.

6.2   *Having a deeper storyline*

Another factor that has a big impact in immersion on the game is the story element. Most RPGs have an element of adventure in them, and a big part of what makes it enjoyable is the *illusion of nonmediation* where *a person fails to perceive or acknowledge the existence of a medium in his or her communication environment and responds as he or she would if the medium were not there* (**doi:10.1111/j.1468-2958.2000.tb00750.x**) A convincing storyline will bring up this feeling and allow a person to feel that they themselves are immersed in the game as they decide on what they will do given the situation.

Out of the three testers, Ian was the only one to suggest for a better and deeper storyline accompanying the game. As mentioned in the observations, Ian said that an interesting story helps in making the game feel a lot more like an RPG.

> "I would recommend it once he [the in-game character] has a main purpose and goal, and the game has a proper story. We should make it a proper RPG." - Ian (Translated into English)

6.3 *Having a right amount of game challenge*

According to (**karat2000user**) people invest time and find fulfillment in mastering a tool to reach a desired goal. By giving the right amount of challenges, players of the game will find satisfaction in the game and will want to keep playing it for a long time. Making the game too easy will make players get bored. On the other hand, making it too difficult, the players will get discouraged and give up playing the game altogether. A good balance of this allows the game to get the players hooked to the game longer.

On the interview, the level of difficulty of the game was asked through a scale of 1 to 10 (10 being the hardest), Dominic gave a score of 7, Philip gave a score of 6, and Ian gave a score of 1. Looking at their responses, Ian gave the lowest score of the three testers. This can be attributed in his fast completion of the game, only taking approximately 5 minutes in total, suggesting there was not much of a challenge for him to play the game. This indicates a possible improvement of the game of adding options for changing its difficulty. This also showed the difference in the three testers' backgrounds in playing mobile games, with experience being an important factor in their perception of the game's difficulty.

6.4 *Using comfortable and intuitive controls*

Comfortable and intuitive controls are also good for games, since they allow users to easily pick up a game and play it. The testers said that the current control scheme of having a virtual joystick to control the character in game is already fine. To further give the players a deeper sense of control, the in-game character's velocity was set to be proportional to the distance between the reference point (where the player initially touches) and the point where they drag the joystick. A minimum and a maximum velocity was also fine-tuned so as to not make the in-game character too fast or too slow. The virtual joystick was also designed so that the player may set a reference point anywhere on the screen, not having to remember any specific portion of the screen for input.

Dominic had a problem of holding the smartphone in a way that may cause false touching (see Figure 3). This can be addressed by explicitly stating how to hold the mobile device. This may be necessary to ensure that the visually-impaired, who does not have visual cues of where they are touching on the screen, will not encounter problems regarding false touch. However, several other issues were encountered during the latter two tests.

While Dominic's interview did not show any issues regarding game controls, Philip and Ian each had their own problems. The main issue which hindered Philip from finishing the game is the screen's automatic rotation, which was due to how he held the device. In the two images below, Figure 5 shows how the device should be held during gameplay, and Figure 4 shows a grip which may cause the device to automatically rotate, messing up the directional orientation of the player. While seemingly a simple issue with the easy fix of just holding the device upright, the lack of proper sound indicators or warnings for events such as this proved to be detrimental to the experience, rendering Philip unable to complete the level.

Philip also made a common gesture while navigating through the levels, which is rapidly and repeatedly swiping the screen towards a specific direction. As mentioned in the research observations, this is perceived as an act of experimentation of trying to make the in-game character walk faster. Since the navigation controls were not designed in such a way and

Figure 4.   Wrong Hold.

Figure 5.   Correct Hold.

repeated gestures do not actually make the in-game character faster, this is a notable indication of a possible alternative control scheme of allowing the character to move faster.

As for Ian, his issue with game controls was the unresponsiveness of the virtual joystick at one point during the test, wherein the device would not recognize his touch input. However, this particular issue was later on determined to be a hardware problem and should not be present using a different device.

The testers made several suggestions for improvement of the game controls before recommending it, specifically, the addition of a main menu wherein vibrations may be turned on or off, locking the screen's orientation to portrait, and adding more and better quality sound assets to deliver information more clearly. Given these improvements, all testers responded that they would recommend the game to others.

All in all, the four (4) themes are applicable when designing and developing RPG games in general, whether they are for visually-impaired or not. In this research, it seems that it is possible to develop a mobile RPG using only audio and haptic components that visually-impaired individuals can enjoy playing.

7   CONCLUSION

Dark Night 2, an RPG, was specifically designed and developed for the visually-impaired individuals. Visual design and development of the game were forgone, and concentration of the design and development were more on capitalizing on sound and haptic game components. A qualitative approach in analyzing the experience of blind users with a mobile role-playing game was chosen because of the limited population. Testing involved observation in a controlled environment of how testers played the game, and an interview determining the effects of the use of the sound and haptic game components on their experience in playing RPGs. Using thematic analysis on data gathered in the observation and interview, the research revealed four (4) major themes in improving the interaction of visually-impaired individuals in playing RPGs. The use of audio as a medium for delivering game information is feasible given that the sound is of high quality and realistic with the right mix. A deeper storyline encourages the player to immerse themselves in the game. A good amount of challenge is also well-appreciated. It should not be too easy that players get bored. It should not be too difficult that they give up. However, how deep the storyline is and how good the challenge is depend on the level of experience of the players in playing games. Having comfortable and intuitive controls also plays an important role in delivering a good game experience, as they could cause unnecessary frustration if the use of the sound and haptic components are not evident and uncomfortable. Keeping the four themes in mind in designing and developing RPGs for the visually-impaired using audio and haptic game elements may lead to an overall enjoyable RPG experience.

ACKNOWLEDGEMENT

The researchers would like to acknowledge *Marithe Girbaud Professorial Chair Award* for making this paper a possibility.

Theory and Practice of Computation – Nishizaki et al (eds)
© 2021 Taylor & Francis Group, London, ISBN 978-0-367-41473-3

Secure remote genome-wide association studies using fully homomorphic encryption

Joey Andrea M. Cruz
Department of Physical Sciences and Mathematics, University of the Philippines Manila, Manila, Philippines

Richard Bryann L. Chua
Department of Physical Sciences and Mathematics, University of the Philippines Manila, Manila, Philippines
Department of Computer Science, University of the Philippines Diliman, Quezon City, Philippines

ABSTRACT: As genomic data become more widely available, the need for powerful computing resources to process large volume of data becomes more critical. As a result, researchers are now increasing the use of cloud computing. However, there is loss of direct control by the researchers upon their data as they outsource computations, leaving the owners of genomic data vulnerable to exploitation and privacy problems. In our work, we explored the use of fully homomorphic encryption through the use of SEAL to securely perform genome-wide association studies remotely.

Keywords: genome privacy, genome-wide association studies, homomorphic encryption, SEAL

1 INTRODUCTION

Since 2005, Genome-Wide Association Studies (GWAS) have shown that the susceptibility of individuals to some diseases are statistically associated to certain sequences found in their DNA. Many of these studies have revealed correlation to previously unsuspected genes and thus have helped formulate new hypotheses for investigation about disease mechanisms and corresponding treatment targets. (Bush & Moore, 2012; Visscher, Brown, McCarthy, & Yang, 2012; Committee et al., 2009; Clarke et al., 2011).

Performing GWAS is now feasible because millions of human DNA sequence variations have been cataloged. New technologies that can assay over one million variants rapidly and accurately have also been developed. With these hight-throughput sequencing technologies, more and more genetic data are becoming available for study. (Bush & Moore, 2012; Visscher et al., 2012; Committee et al., 2009; Clarke et al., 2011).

Researchers are now using cloud computing more for their resources, as it can better accommodate the great volume of data now readily accessible. A very important consideration, however, is the loss of direct control by the researchers upon their data as they outsource computations. While this should greatly accelerate the processing of data, it leaves these very sensitive genomic data, and consequently its donors, vulnerable to exploitation. (*Creating a Global Alliance to Enable Responsible Sharing of Genomic and Clinical Data, 2013*) In fact, it has been shown that even with only a small subset of the genome, 30-80 SNPs out of millions, it is possible to infer the identity of an individual, and estimate his/her disease risk (Lin, Owen, & Altman, 2004). Consequently, by the nature of genomic data, one might infer information about the individual's family (Humbert, Ayday, Hubaux, & Telenti, 2013). Thus, the malicious use of these data exposes the individuals, and even their close kin, to possible exploitation by discrimination due to their genomic predisposition. Several studies have shown that limiting to only the removal of the personal information does not solve the privacy problem, as the genomic data themselves can be used to infer a person's identity. Numerous works identify approaches that illustrate how such threats to privacy can be realized. Even when the data are anonymized or de-identified (i.e. explicit identifiers are removed), identity information can still be inferred as the data themselves are an individual's identity code. The standard anonymization/de-identification approach to privacy is no longer deemed applicable to genomic data (Kamm, Bogdanov, Laur, & Vilo, 2013). For example, there is already an approach for re-identification using inferred phenotypes from public data and a list of known associations between genotype and phenotypic traits (Malin & Sweeney, 2001). This can be used by malicious entities, and can become more practical as the list of associations grow with further studies (Hindorff et al., 2009). Furthermore, the threat of re-identifying anonymized genotype data is compounded by the size of online genotype repositories now available (Gymrek, McGuire, Golan, Halperin, & Erlich, 2013). Aggregated pools of genomic data, including GWAS statistics, can also leak private information. For GWAS, specifically, the presence of a participant in a case group can be inferred using statistical methods from the group's aggregated genomic data (Homer et al., 2008; Craig et al., 2011; Gymrek et al., 2013; Cai et al., 2015). The use of genomic data-sharing beacons also present privacy risks for genomic data donors. These beacons are web servers that answer allele-presence queries—i.e. whether the beacon contains a certain allele in a specified location. It is shown by Shringarpure and Bustamante (Shringarpure & Bustamante, 2015) that through statistical methods, the presence of an individual's genome in a beacon can be inferred with significant power by performing multiple queries. The number of queries necessary depends on the number of individuals in the beacon, given the desired statistical power.

With all of these threats, there is a need to securely outsource GWAS computations. Ideally, we want to be able to perform GWAS computations remotely without the need to upload the plain genomic data in order to minimize the privacy risk. Homomorphic encryption is an ideal solution to this problem. In our work, we used fully homomorphic encryption to perform some GWAS computations. In section 2, we did a review of the different works on securely performing GWAS computations remotely. In sections 3 and 4, we briefly introduced homomorphic encryption and the SEAL library, respectively. We then discussed how we used SEAL in developing a solution to securely outsource GWAS computation in section 5.

2 LITERATURE REVIEW

Differential privacy, secure multi-party computation, and cryptography are among the most widely used techniques in providing security for genomic data. Simmons et al. (Simmons, Sahinalp, & Berger, 2016), Yu et al. (Yu, Fienberg, Slavkovi´c, & Uhler, 2014), and Tramer et al. (Tram`er, Huang, Hubaux, & Ayday, 2015) used the concept of differential privacy, a form of data perturbation, in genome-wide association studies (GWAS). Differential privacy guarantees that an analysis performed on any dataset is statistically indistinguishable from the same analysis performed on any dataset that differs in any individual's disease status. Zhang et al. (Zhang, Blanton, & Almashaqbeh, 2015) and Kamm et al. (Kamm et al., 2013) used secure multi-party computation in their works. Both use the mechanism of secret sharing, which ensures that each computing party gets a subset of the data that leaks no information on its own, but is meaningful when the data are pooled back together.

Works by Wang et al. (Wang et al., 2016), Lauter et al. (Lauter, L´opez-Alt, & Naehrig, 2014), Lu et al. (Lu, Yamada, & Sakuma, 2015), Kim and Lauter (Kim & Lauter, 2015), and Zhang et al. (Zhang, Dai, Jiang, Xiong, & Wang, 2015) used homomorphic encryption in GWAS. Fully homomorphic encryption (FHE) enables meaningful computation on data without knowledge of the secret key. In these works, all genomic data in the form of SNPs are encoded and encrypted locally before they are uploaded to the cloud which then conducts computations using homomorphic operations and returns encrypted results. Lu et al. supported the evaluation of the D' measure of linkage disequilibrium, Hardy-Weinberg equilibrium, and the χ^2 test statistic using the BGV scheme in their approach. A packing technique was used to represent vectors of integers, which condenses the sequence of homomorphic additions and multiplications involved in the computation of a scalar product into a single homomorphic multiplication. The use of this packing technique effectively decreased runtime, as it reduced the number of ciphertexts necessary to represent a SNP and the number of homomorphic operations required.

3 HOMOMORPHIC ENCRYPTION

Homomorphic encryption refers to a class of encryption schemes that allow computation on ciphertexts that will decrypt to the result of computing on the original plaintexts. With homomorphic encryption, we are able to perform meaningful computations on ciphertext remotely without the need of decrypting them. Formally, it is defined as follows: Let M denote the set of the plaintexts and C denote the set of ciphertexts. An encryption scheme is said to be homomorphic if for any given public-private key pair (but we omitted the public-private key pair arguments in the encryption and decryption functions to make them shorter) the encryption function E and the decryption function D satisfies

$$D(E(m_1) \underset{C}{\odot} E(m_2)) = m_1 \underset{M}{\odot} m_2$$

for some operators $\odot_M$ in M, $\odot_C$ in C and for all m_1, m_2 in M. (Aslett, Esperanc,a, & Holmes, 2015)

A scheme is additively homHence, all elements of this ring can be expressedomorphic if the addition operator is homomorphic, while it is multiplicatively homomorphic if the multiplication operator is homomorphic. A scheme is said to be partially homomorphic if it is only homomorphic on one operator and this operator can be performed with unlimited number of times. A scheme is said to be somewhat homomorphic if it is both additively and multiplicatively homomorphic but these operators can be performed with limited number of times. A scheme is said to be fully homomorphic if it is both additively and multiplicatively homomorphic and these operators can be performed unlimited number of times. Somewhat and fully homomoprhic schemes are of interest in the field of homomorphic encryption as the addition and multiplication operations allow us to build arbitrary functions. (Morris, 2013; Acar, Aksu, Uluagac, & Conti, 2018)

4 SIMPLE ENCRYPTED ARITHMETIC LIBRARY (SEAL)

Simple Encrypted Arithmetic Library (SEAL) is a homomorphic encryption solution made publicly available by Microsoft Research (Chen, Laine, & Player, 2016). SEAL is written in C ++ and comes with a C# wrapper library called SEALNET. It is an implementation of the Fan-Vercauteren (FV) scheme (Fan & Vercauteren, 2012), a leveled homomorphic encryption scheme, and consists of key generation, encryption, decryption, homomorphic addition, and homomorphic multiplication algorithms. The scheme operates in the ring of polynomials with integer coefficients of a degree less than n. Hence, all elements of this ring can be expressed in the form $\sum_{i=0}^{n-1} a_i x^i$ where $a_i \in Z$.

Together with HELib, created by IBM and implemented the Brakerski-Gentry-Vaikuntanathan (BGV) scheme, these are the two widely-used fully homomorphic encryption libraries. We chose to use SEAL over HELib because of the better documentation that comes with SEAL. Morevover, SEAL has a helper class, ChooserEvaluator, which made the selection of parameters easier. Selecting the correct parameters in homomorphic encryption that strikes a balance between correctness by minimizing the noise and computational efficiency is a difficult task (Chen et al., 2016).

In all proceeding discussion, when we say polynomial, we mean by it a polynomial that is a member of the ring discussed above. In the FV scheme, a plaintext is a polynomial and a freshly encrypted ciphertext is an array of two polynomials. The secret key is a polynomial, and the public key is a polynomial array of size two. Addition and multiplication may be performed over two ciphertexts or a ciphertext and a plaintext.

5 USING SEAL FOR GWAS

5.1 Problem setting

We consider the problem setting where there are two stakeholders: researcher (which acts as the client) and a cloud server (which performs the GWAS computation). The researcher has all the SNP data and he/she uploads it to the cloud server, where the GWAS computations are performed homomorphically. We use the sample data provided in challenge 1 of the iDASH Privacy and Security Workshop 2015 Secure Genome Analysis Competition.[1]

A sample input data is provided in figure 1. The first line in the file contains the identifiers for each of the genotype samples in the file. The proceeding lines are all of SNP data, and each SNP is represented by two lines. The first row of each set of SNPs is the RSID of the SNP, which uniquely identifies the SNP locus. The second row of each set of SNPs enumerates the genotype occurrences for each sample. There are, in this line, as many genotypes as there are identifiers in the first line of the file, and each of the genotypes enumerated correspond to the identifiers in the order they were given.

5.2 Homomorphic computation of χ^2 test statistics, Allelic Odds ratio, Minor Allele frequency, and Hardy-Weinberg equilibrium

Since the plaintext space of FV scheme is a polynomial, we have to encode our SNP data into polynomial representation. We used the encoding technique used by Lu et. al. in (Lu et al., 2015), since they used the BGV scheme which also has a plaintext space that is polynomial. The SNP data are encoded into plaintext polynomials, as follows: Assuming a biallelic locus of, say, alleles A and B, each SNP is represented by four plaintext polynomials, ρ_{fw}^{AA}, ρ_{bw}^{AA}, ρ_{fw}^{AB}, and ρ_{bw}^{AB}. ρ_{fw}^{AA} and ρ_{fw}^{AB} are called forward-packed polynomials; ρ_{bw}^{AA} and ρ_{bw}^{AB} are called backward-packed polynomials. Each of these polynomials are of degree $M - 1$, where M is the number of genotype samples given.

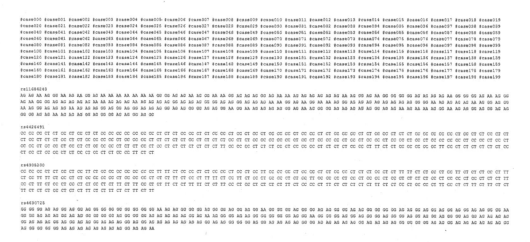

Figure 1. Sample text file containing SNP data in iDASH format.

1. http://www.humangenomeprivacy.org/2015/competition-tasks.html

The forward-packed polynomials can be expressed as $\rho_{fw}^{X} = \sum_{i=1}^{M} a_i x^{i-1}$, $X \in \{AA, AB\}$ where a_i is the frequency of the allele A at the ith genotype, if the ith genotype is X. It is zero otherwise. The backward-packed polynomials can be expressed as $\rho_{bw}^{X} = \sum_{i=1}^{M} a_i x^{M-i}$, $X \in \{AA, AB\}$, where the a_i's are defined similarly as that of the forward-packed polynomial. To illustrate this, take a biallelic locus associated with the alleles A and G. The genotype data AA, AG, AG, AA, GG are encoded as follows:

$$\rho_{fw}^{AA} = 2x^0 + 0x^1 + 0x^2 + 2x^3 + 0x^4, \quad \rho_{bw}^{AA} = 0x^0 + 2x^1 + 0x^2 + 0x^3 + 2x^4,$$

$$\rho_{fw}^{AG} = 0x^0 + 1x^1 + 1x^2 + 0x^3 + 0x^4, \quad \rho_{bw}^{AG} = 0x^0 + 0x^1 + 1x^2 + 1x^3 + 0x^4.$$

The case-control membership statuses of the subjects are also encoded in a polynomial of degree $M - 1$. This polynomial is defined as $\gamma^{case} = \sum_{i=1}^{M} a_i x^{M-i}$, where a_i is one if the ith subject is in the case group and zero if he/she is in the control group. Here, we note that the construction of γ^{case} is similar to that of a backward-packed polynomial. Hence, in the proceeding discussions, we also use it as such. For example, if five subjects have the statuses {case, case, case, control, control}, then the corresponding γ^{case} will be $0x^0 + 0x^1 + 1x^2 + 1x^3 + 1x^4$. We used this type of encoding in order to facilitate the computation of the scalar product of two vectors, which is the middle term of the product of the two polynomials. We refer readers to (Lu et al., 2015) for the proof that this is the case. This product is used to find the frequency of an allele/genotype that conincides with particular conditions. To illustrate, the number of homozygous genotypes for the reference allele is the coefficient of the middle term of the polynomial that results from multiplying ρ_{fw}^{AA} with γ^{case}. In the case that no restricting condition is used, the desired frequency can be computed by multiplying the forward packed polynomial to a polynomial of equal degree whose coefficients are all equal to one. Hence, the total number of homozygous genotypes for the reference allele across both the case and control groups is the coefficient of the middle term of the polynomial that results from multiplying ρ_{fw}^{AA} with the polynomial of coefficients equal to one.

These polynomial encodings of SNP data are then encrypted and uploaded to the remote server where the computations are performed homomorphically. Selection of encryption parameters are based on the size of the input and the type of computation to be performed. The encryption parameters have to be selected correctly to ensure that the homomorphic computations evaluates correctly – i.e. that the noise introduced in the encryption will not cause the homomorphic computations to fail. In SEAL, these parameters are the polynomial modulus poly_mod, the coefficient modulus coeff_mod, and the plaintext modulus plain_mod. In our computation for this, we set poly_mod as $x^{2^{(\log_2 M+2)}} + 1$ and plain_mod as $2^{(\log_2 M+2)}$. The coeff_mod was set to the suggested values by SEAL which is based on security given by the parameter poly_mod.

The computation of the χ^2 test statistic, allelic odds ratio, minor allele frequency, and Hardy-Weinberg equilibrium all involve frequency data from an allelic contingency table (Table 1) (Kim & Lauter, 2015).

This contingency table can be constructed with the knowledge of the size of the entire case-control group—i.e. the number of subjects, which we denote by M—and the three frequencies N_A^{case}, N_A, and N^{case} derived from the genomic data. It is sufficient to know these four frequencies from the genomic data in order to compute the mentioned statistics.

Let $a_i(\rho)$ be the ith coefficient of the polynomial ρ, and P_n be the polynomial of degree $n - 1$ with all coefficients equal to 1. With the encoding scheme we used, the frequencies N_A^{case} and N_A can be computed as follows

$$N_A^{case} = a_M \left(D \left(\left(E \left(\rho_{fw}^{AA} \right) + E \left(\rho_{fw}^{AB} \right) \right) E(\gamma^{case}) \right) \right),$$

$$N_A = a_M \left(D \left(\left(E \left(\rho_{fw}^{AA} \right) + E \left(\rho_{fw}^{AB} \right) \right) E(P_M) \right) \right),$$

Table 1. Allelic contingency table.

	Allele Type		
	A	B	Total
Case	N_A^{case}	N_B^{case}	N^{case}
Control	$N_A^{control}$	$N_B^{control}$	$N^{control}$
Total	N_A	N_B	

The resulting encrypted polynomials are returned to the client for decryption and construction of the allelic contingency table using the frequencies given by the middle coefficients of the decrypted polynomials. The remaining necessary quantities are computed as follows:

$$N_B^{case} = 2N^{case} - N_A^{case}$$

$$N_A^{control} = N_A - N_A^{case}$$

$$N_B = 2M - N_A$$

$$N^{control} = M - N^{case}$$

$$N_B^{control} = N_B - N_B^{case} = N^{control} - N_A^{control}$$

After the construction of the allelic contingency table, the χ^2 test statistic, allelic odds ratio, and minor allele frequency can be computed directly using their respective formulas, as given in Table 2. Hardy-Weinberg equilibrium computation additionally requires the genotype frequencies N_{AA}, N_{AB}, and N_{BB}. N_{AA} and N_{AB} can be computed remotely and similarly returned to the client as

$$2N_{AA} = a_M\left(D\left(E\left(\rho_{fw}^{AA}\right)E(P_M)\right)\right),$$

$$N_{AB} = a_M\left(D\left(E\left(\rho_{fw}^{AB}\right)E(P_M)\right)\right).$$

N_{BB} is then given by $M - N_{AA} - N_{AB}$.

Table 2. Formulas for allelic case-control contingency table-based statistics.

Statistic	Formula
χ^2	$\dfrac{2M(N_B^{case}(N_A^{control}+N_B^{control})-N_B^{control}(N_A^{case}+N_B^{case}))^2}{N^{case}N^{control}N_A N_B}$
AOR	$\dfrac{N_A^{case}N_B^{control}}{N_A^{control}N_B^{case}}$
MAF	$\dfrac{\min(N_A,N_B)}{N_A+N_B}$
HWE	$\sum_{X\in\{AA,AB,BB\}}\dfrac{(N_X-E_X)^2}{E_X}$

Table 3.   Genotypic frequency table.

| | | Marker M_1 | | | |
		AA	Aa	aa	Total
	BB	N_{AABB}	N_{AaBB}	N_{aaBB}	N_{BB}
Marker M_2	Bb	N_{AABb}	N_{AaBb}	N_{aaBb}	N_{Bb}
	bb	N_{AAbb}	N_{Aabb}	N_{aabb}	N_{bb}
	Total	N_{AA}	N_{Aa}	N_{aa}	$2M$

5.3   Homomorphic computation of linkage disequilibrium and heterozygosity rate

To compute for the linkage disequilibrium and heterozygosity rate, we need the genotypic frequency table (Tabe 3) (Lu et al., 2015).

For two biallelic loci, the first with reference allele A and alternate allele a, and the second with reference allele B and alternate allele b, linkage disequilibrium is defined as

$$D' = \frac{D}{D_{max}},$$

where

$$D = pAB - pApB,$$

and

$$D_{max} = \begin{pmatrix} \min\{pA(1 - pB), (1 - pA)pB\} & ,D > 0 \\ \min\{pApB, (1 - pA)(1 - pB)\} & ,D < 0 \end{pmatrix}.$$

The quantities necessary to complete this computation are the proportions p_{AB}, p_A, and p_B. These can be computed as

$$p_{AB} = \frac{2N_{AABB} + N_{AaBB} + N_{AABb}}{2M}$$

$$p_A = \frac{2N_{AA} + N_{Aa} - N_{AaBb}}{2M}$$

$$p_B = \frac{2N_{BB} + N_{Bb} - N_{AaBb}}{2M}.$$

These required computations can be further broken down in terms of the quantities N_{AABB}, N_{AaBB}, N_{AABb}, N_{AaBb}, $2N_{AA} + N_{Aa}$, and $2N_{BB} + N_{Bb}$. These, in turn, can be computed homomorphically as

$$4N_{AABB} = a_M\left(D\left(E\left(\rho_{fw}^{AA}\right)E(\rho_{bw}^{BB})\right)\right)$$

$$2N_{AaBB} = a_M\left(D\left(E\left(\rho_{fw}^{Aa}\right)E(\rho_{bw}^{BB})\right)\right)$$

$$2N_{AABb} = a_M\left(D\left(E\left(\rho_{fw}^{AA}\right)E(\rho_{bw}^{Bb})\right)\right)$$

$$N_{AaBb} = a_M\left(D\left(E\left(\rho_{fw}^{Aa}\right)E(\rho_{bw}^{Bb})\right)\right)$$

$$2N_{AA} + N_{Aa} = a_M\left(D\left(\left(E\left(\rho_{fw}^{AA}\right) + E\left(\rho_{fw}^{Aa}\right)\right)E(P_M)\right)\right)$$

$$2N_{BB} + N_{Bb} = a_M\left(D\left(\left(E\left(\rho_{fw}^{BB}\right) + E\left(\rho_{fw}^{Bb}\right)\right)E(P_M)\right)\right)$$

With this method, we can compute for the linkage disequililbrium of the loci pairs consisting of the first SNP in the input file and each of the proceeding SNPs given.

The total heterozygosity across N SNPs for several individuals can be computed homomorphically as

$$E(\rho_{HR}) = \sum_{i=1}^{N} E\left(\rho_{fw}^{Aa}\right).$$

The heterozygosity rate of the ith individual is given by

$$\frac{a_i(D(E(\rho_{HR})))}{N}$$

For linkage disequilibrium, the SEAL parameters we used are $x^{2^{(\log_2 M+3)}} + 1$ for poly_mod and $2^{(\log_2 M+4)}$ for plain_mod; for heterozygosity rate, we used $x^{2^{(\log_2 M+2)}} + 1$ for poly_mod and $2^{(\log_2 M+1)}$ for plain_mod.

5.4 *Performance evaluation*

To assess the performance of our homomorphic computation using SEAL, we measured their running times with the following environment: Intel Core i7-4510u @ 2.60GHz, 8GB RAM, running 64-bit Windows 10. We took the following time measurements:

1. TTE – Total time elapsed from the beginning of the computation to its completion (including transmission time)
2. AET – Average encryption time
3. ADT – Average decryption time
4. ART – Average remote computation time (i.e. covers the server computation time and transmission time)

In the computation of the χ^2 test statistic, allelic odds ratio, Hardy-Weinberg equilibrium, minor allele frequency, and linkage disequilibrium, an input corresponds to a sequence of identical transactions to the server. That is, for each SNP in the file, the same computation occurs and these computations are independent of each other. Hence, we could parallelize the computation with multithreading in both the client and server side.

Tables 4, 5, and 6 contain the running times for single-threaded, double-threaded, and triple-threaded implementations operating upon 25 SNPs. The TTE decreases as the number of threads increases. In comparison to the single-threaded implementation, double-threading causes an average of 1.6 speedup while triple-threading can cause an average of 2.3 speedup.

Table 4. Running times for one thread with 25 SNPs.

Statistic	TTE	AET	ADT	ART
χ^2 test statistic	233 s	344 ms	446 ms	7974 ms
allelic odds ratio	233 s	344 ms	446 ms	7970 ms
Hardy-Weinberg equilibrium	280 s	343 ms	442 ms	9820 ms
minor allele frequency	174 s	345 ms	376 ms	5676 ms
linkage disequilibrium	564 s	714 ms	671 ms	22704 ms

Table 5. Running times for two threads with 25 SNPs.

Statistic	TTE	AET	ADT	ART
χ^2 test statistic	146 s	367 ms	482 ms	8004 ms
allelic odds ratio	146 s	364 ms	488 ms	8009 ms
Hardy-Weinberg equilibrium	173 s	376 ms	484 ms	9859 ms
minor allele frequency	118 s	407 ms	420 ms	5712 ms
linkage disequilibrium	306 s	771 ms	746 ms	22790 ms

Table 6. Running times for three threads with 25 SNPs.

Statistic	TTE	AET	ADT	ART
χ^2 test statistic	101 s	395 ms	501 ms	8056 ms
allelic odds ratio	102 s	402 ms	509 ms	8080 ms
Hardy-Weinberg equilibrium	122 s	449 ms	541 ms	9920 ms
minor allele frequency	181 s	430 ms	450 ms	5761 ms
linkage disequilibrium	210 s	842 ms	751 ms	23223 ms

AET, ADT, and ART all increase as the number of threads increases. This is to be expected as a result of executing computations in parallel.

In light of this observation, the proceeding discussions use the triple-threaded implementation because while the AET, ADT, and ART increase, the TTE still is minimum after parallelization of computation.

Tables 7, 8, 9, 10, 11, and 12 contain the average running times for the computation of each of the statistics for 10, 25, 50, 100, and 500 SNPs, respectively, each with three replications.

In the tabulation of these results, AET, ADT, and ART all remain around the same range regardless of how many SNPs are being processed in any of the computations. The TTE is observed to increase linearly with the number of SNPs, as expected, because despite the use of three threads, the computation eventually leads into a sequence of three computations occurring in parallel.

Table 7. Average running times for the computation of χ^2 test statistic.

	TTE	AET	ADT	ART
10 SNPs	46 s	399 ms	508 ms	8119 ms
25 SNPs	102 s	411 ms	500 ms	8076 ms
50 SNPs	196 s	413 ms	503 ms	8068 ms
100 SNPs	407 s	417 ms	513 ms	8080 ms
500 SNPs	1908 s	395 ms	514 ms	8032 ms

Table 8. Average running times for the computation of allelic odds ratio.

	TTE	AET	ADT	ART
10 SNPs	46 s	410 ms	523 ms	8089 ms
25 SNPs	102 s	411 ms	502 ms	8078 ms
50 SNPs	196 s	420 ms	516 ms	8077 ms
100 SNPs	406 s	411 ms	515 ms	8065 ms
500 SNPs	1910 s	405 ms	517 ms	8034 ms

Table 9. Average running times for the computation of Hardy-Weinberg equilibrium.

	TTE	AET	ADT	ART
10 SNPs	54 s	410 ms	524 ms	9989 ms
25 SNPs	122 s	411 ms	524 ms	9932 ms
50 SNPs	228 s	420 ms	501 ms	9898 ms
100 SNPs	452 s	411 ms	501 ms	9893 ms
500 SNPs	2279 s	434 ms	524 ms	9866 ms

Table 10. Average running times for the computation of minor allele frequency.

	TTE	AET	ADT	ART
10 SNPs	36 s	409 ms	446 ms	5731 ms
25 SNPs	82 s	431 ms	443 ms	5728 ms
50 SNPs	153 s	417 ms	419 ms	5719 ms
100 SNPs	303 s	418 ms	430 ms	5718 ms
500 SNPs	1483 s	419 ms	414 ms	5717 ms

Table 11. Average running times for the computation of linkage disequilibrium.

	TTE	AET	ADT	ART
10 SNPs	79 s	823 ms	759 ms	22366 ms
25 SNPs	208 s	838 ms	741 ms	22669 ms
50 SNPs	435 s	816 ms	748 ms	22599 ms
100 SNPs	860 s	823 ms	753 ms	22297 ms
500 SNPs	4262 s	810 ms	748 ms	22551 ms

Table 12. Average running times for the computation of heterozygosity rate.

	TTE	AET	ADT	ART
10 SNPs	17 s	10 ms	4 ms	348 ms
25 SNPs	42 s	19 ms	4 ms	851 ms
50 SNPs	83 s	34 ms	4 ms	1690 ms
100 SNPs	165 s	64 ms	4 ms	3369 ms
500 SNPs	821 s	314 ms	4 ms	16796 ms

With all timings considered, we observed that the AET, ADT, and ART change with the number of threads used, increasing as the number of threads do and the TTE decreases as the number or threads increase and increases as the number of SNPs increase. The trend in computation times (TTE) with respect to the amount of threading suggests that the server's capacity to parallelize computation should make computations of statistics across multiple SNPs occur at a reasonable duration. However, we observed increases in the averaged duration measurements and these likely can be attributed to the setup cost of the threads in both client (AET and ADT) and the server (ART).

6 CONCLUSIONS AND FUTURE WORK

In our work, we showed how to use fully homomorphic encryption through the use of SEAL in securely computing basic GWAS statistics like the χ^2 test statistic, allelic odds ratio, Hardy-Weinberg equilibrium, minor allele frequency, linkage disequilibrium, and heterozygosity rate. Through the use of SEAL, it allowed us to better select the optimal security parameters in order to get the correct result of the computations. During the time of our development, SEAL 2.0 is the latest version of SEAL available. In 2018, SEAL 3.0 was released that implements BFV and Cheon-Kim-Kim-Song (CKKS) schemes. We plan to look if SEAL 3.2.0 (released on 21 February 2019) would give us a better development experience and improved efficiency. One of the advantages of the CKKS scheme is the ability to perform homomorphic computations on encrypted real and complex numbers by approximating floating-point computations. In our current work, any division that yields real number is performed locally. We would like to see if with CKKS, we could possibly perform all computations on the server. Further enhancements can be made by supporting the computation of other statistics used in GWAS especially those that deal with real numbers and the handling of violations of some assumptions such as biallelism of SNP loci and completeness of data.

To this date, there are already at least 10 open source HE libraries available (Hallman et al., 2018). Although SEAL and HELib still remain to be the two widely-used HE libraries, we could look at the other HE libraries and see if they can offer better performance. In fact, PALISADE has already been applied to GWAS (Blatt, Gusev, Polyakov, Rohloff, & Vaikuntanathan, 2019). We could also look into how GPU and FPGA could be used to improved the performance of HE operations. Some of the HE libraries are already supporting GPUs.

REFERENCES

Acar, A., Aksu, H., Uluagac, A. S., and Conti, M. (2018, jul). A Survey on Homomorphic Encryption Schemes. *ACM Computing Surveys*, 51(4):1–35. Retrieved from http://dl.acm.org/citation.cfm?doid=3236632.3214303 doi:10.1145/3214303

Aslett, L. J., Esperança, P. M., & Holmes, C. C. (2015). *A review of homomorphic encryption and software tools for encrypted statistical machine learning*.

Blatt, M., Gusev, A., Polyakov, Y., Rohloff, K., and Vaikuntanathan, V. (2019). *Optimized homomorphic encryption solution for secure genome-wide association studies*. Cryptology ePrint Archive, Report 2019/223 (https://eprint.iacr.org/2019/223)

Bush, W. S. and Moore, J. H. (2012). Genome-wide association studies. *PLoS Computational Biology*, 8(12).

Cai, R., Hao, Z., Winslett, M., Xiao, X., Yang, Y., Zhang, Z., and Zhou, S. (2015). Deterministic identification of specific individuals from GWAS results. *Bioinformatics*, 31(11), 1701–1707.

Chen, H., Laine, K., and Player, R. (2016). *Simple encrypted arithmetic library - seal (v2.1)*. Technical report. Retrieved from https://www.microsoft.com/en-us/esearch/publication/simple-encrypted-arithmetic-library-seal-v2-1/

Clarke, G. M., Anderson, C. A., Pettersson, F. H., Cardon, L. R., Morris, A. P., & Zondervan, K. T. (2011). Basic statistical analysis in genetic case-control studies. *Nature Protocols*, 6(2), 121–133.

Committee, P. G. C. C. et al. (2009). Genome wide association studies: history, rationale, and prospects for psychiatric disorders. *American Journal of Psychiatry*.

Craig, D. W., Goor, R. M., Wang, Z., Paschall, J., Ostell, J., Feolo, M.,... Manolio, T. A. (2011). Assessing and managing risk when sharing aggregate genetic variant data. *Nature Reviews Genetics*, 12(10), 730–736.

Creating a global alliance to enable responsible sharing of genomic and clinical data. (2013). http://genomicsandhealth.org/about-the-global-alliance/key-documents/white-paper-creating-global-alliance-enable-responsible-shar. (Accessed: 2016-11-02)

Fan, J. and Vercauteren, F. (2012). Somewhat practical fully homomorphic encryption. *IACR Cryptology ePrint Archive*, 2012,144.

Gymrek, M., McGuire, A. L., Golan, D., Halperin, E., and Erlich, Y. (2013). Identifying personal genomes by surname inference. *Science*, 339(6117), 321–324.

Hallman, R., Laine, K., Dai, W., Gama, N., Malozemoff, A., Polyakov, Y., and Carpov, S. (2018). *Building applications with homomorphic encryption, a presentation from the homomorphic encryption standardization consortium*. http://homomorphicencryption.org/wp-content/uploads/2018/10/CCS-HE-Tutorial-Slides.pdf. (Accessed: 2018-12-01)

Hindorff, L. A., Sethupathy, P., Junkins, H. A., Ramos, E. M., Mehta, J. P., Collins, F. S., and Manolio, T. A. (2009). Potential etiologic and functional implications of genome-wide association loci for human diseases and traits. *Proceedings of the National Academy of Sciences, 106*(23), 9362–9367.

Homer, N., Szelinger, S., Redman, M., Duggan, D., Tembe, W., Muehling, J., Pearson, J. V., Stephan, D. A., Nelson, S. F., and Craig, D. W. (2008). Resolving individuals contributing trace amounts of dna to highly complex mixtures using high-density snp genotyping microarrays. *PLoS Genetics, 4*(8), e1000167.

Humbert, M., Ayday, E., Hubaux, J.-P., and Telenti, A. (2013). Addressing the concerns of the lacks family: quantification of kin genomic privacy. In *Proceedings of the 2013 acm SIGSAC Conference on Computer & Communications Security*, (pp. 1141–1152).

Kamm, L., Bogdanov, D., Laur, S., and Vilo, J. (2013). A new way to protect privacy in large-scale genome-wide association studies. *Bioinformatics (Oxford, England), 29*(7), 886–93. Retrieved from http://www.pubmedcentral.nih.gov/articlerender.fcgi?artid=3605601{\&}tool=pmcentrez{\&}rendertype=abstractdoi:10.1093/bioinformatics/btt066

Kim, M. and Lauter, K. (2015). Private genome analysis through homomorphic encryption. *BMC Medical Informatics and Decision Making, 15*(Suppl 5), S3.

Lauter, K., López-Alt, A., & Naehrig, M. (2014). Private computation on encrypted genomic data. In *International Conference on Cryptology and Information Security in Latin America*, 3–27. Springer.

Lin, Z., Owen, A. B., & Altman, R. B. (2004). Genomic research and human subject privacy. *Science, 305*(5681).

Lu, W.-J., Yamada, Y., and Sakuma, J. (2015). Privacy-preserving genome-wide association studies on cloud environment using fully homomorphic encryption. *BMC Medical Informatics and Decision Making, 15*(Suppl 5), 1–8.

Malin, B. and Sweeney, L. (2001). Inferring genotype from clinical phenotype through a knowledge based algorithm. In *Proceedings of the Pacific Symposium on Biocomputing*, (pp 41–52).

Morris, L. (2013). Analysis of partially and fully homomorphic encryption. http://www.liammorris.com/crypto2/Homomorphic%20Encryption%20Paper.pdf. (Accessed: 2016-11-02)

Shringarpure, S. S. and Bustamante, C. D. (2015). Privacy risks from genomic data-sharing beacons. *The American Journal of Human Genetics, 97*(5), 631–646.

Simmons, S., Sahinalp, C., & Berger, B. (2016). Enabling privacy-preserving GWASs in heterogeneous human populations. *Cell Systems, 3*(1), 54–61.

Tramèr, F., Huang, Z., Hubaux, J.-P., & Ayday, E. (2015). Differential privacy with bounded priors: reconciling utility and privacy in genome-wide association studies. In *Proceedings of the 22nd acm SIGSAC Conference on Computer and Communications Security*, (pp. 1286–1297).

Visscher, P. M., Brown, M. A., McCarthy, M. I., & Yang, J. (2012). Five years of gwas discovery. *The American Journal of Human Genetics, 90*(1), 7–24.

Wang, S., Zhang, Y., Dai, W., Lauter, K., Kim, M., Tang, Y., Xiong, H., & Jiang, X. (2016). Healer: Homomorphic computation of exact logistic regression for secure rare disease variants analysis in gwas. *Bioinformatics, 32*(2), 211–218.

Yu, F., Fienberg, S. E., Slavkovi, A. B., & Uhler, C. (2014). Scalable privacy-preserving data sharing methodology for genome-wide association studies. *Journal of Biomedical Informatics, 50*,133–141.

Zhang, Y., Blanton, M., & Almashaqbeh, G. (2015a). Secure distributed genome analysis for GWAS and sequence comparison computation. *BMC Medical Informatics and Decision Making, 15*(Suppl 5), S4.

Zhang, Y., Dai, W., Jiang, X., Xiong, H., & Wang, S. (2015b). Foresee: Fully outsourced secure genome study based on homomorphic encryption. *BMC medical informatics and decision making, 15*(Suppl 5), S5.

Theory and Practice of Computation – Nishizaki et al (eds)
© 2021 Taylor & Francis Group, London, ISBN 978-0-367-41473-3

SugarTraces: A persuasive technology-enabled mobile application for diabetics

R.B. Austria, D.C. Caingat, R.P. Feria, L.L. Figueroa & Ma. R. Solamo
College of Engineering, Department of Computer Science, University of the Philippines Diliman, Philippines

ABSTRACT: Medical practitioners around the world use the aid of mobile health (mHealth) technology to provide better, efficient, and convenient health care. With this advantage of mHealth technology, different health applications targeting various diseases were developed. For instance, there are numerous mHealth applications specifically made for patients diagnosed with diabetes. Majority of these provide the standard functionality of blood glucose monitoring. However, persuasion to keep healthy is not a major concern for developers. Apart from recording, these applications lack behavior change techniques used to encourage the user to develop better health habits and lifestyle.

SugarTraces is a motivational mobile health application designed with persuasive components such as feedback and reward system. The feedback system is implemented to motivate the user to continuously use the application while maintaining healthy blood glucose levels. The streaks, achievements, and feedback messages are applied to add gamified and persuasive elements to the application.

SugarTraces was tested by forty-five (45) users over a period of at least five (5) days. The focus of the testing phase was to determine the participants' impression of the feedback messages and the different gamified elements in the application. General overall result is that majority of the users agree that SugarTraces can help manage diabetes. It has an overall score of 4.64 out of 5. Meanwhile, keeping a streak has an overall score of 3.91 out of 5 due to users initially not knowing what it is for.

1 INTRODUCTION

Diabetes is a chronic disease that occurs either when the pancreas does not produce enough insulin or when the body cannot effectively use the insulin it produces. Insulin is a hormone that regulates blood glucose [21]. Meanwhile, hyperglycemia is known as raised blood glucose. It is an effect of uncontrolled diabetes and can cause serious damage to the body.

In the Philippines, diabetes is one of the main causes of mortality [13]. It was the 6th leading cause of death among Filipinos in 2013 according to Philippine Health Statistics and as declared by Philippine Center for Diabetes Education Foundation in 2016. Over 6 million Filipinos are already diagnosed with diabetes. [12]

In 2016, a total of 582,183 deaths were recorded by the Philippine Statistics Authority (PSA). Among these, 74,134 (12.7%) were caused by ischaemic heart diseases. This was the leading cause of mortality in 2016. Meanwhile, diabetes mellitus comes in 6th with 33,295 (5.7%). [5]

With this, a change of lifestyle is needed to help reduce the number of patients affected by diseases rooted from high blood glucose levels. Increasing awareness and urging people to follow a healthy lifestyle are important. Furthermore, healthy diet, regular exercise, maintenance of normal body weight and self-management are as important as medication. [20] Self-management which include tracking of health data such as weight, physical activity, blood pressure, blood glucose level, etc. is more personal and involves less interaction with medical facilities and practitioners. This can be achieved through the use of smart phones and mobile applications with persuasive components.

The use of smart phone and mobile applications affects the daily lives of the population today. In a survey published in 2016, 65.3% of the population in the Philippines used mobile phones in 2015. It increased from 63.2% in 2014 and is expected to continuously increase to 69% in 2019-2020. [2]

The main use of smart phones to consumers are chatting, social networking, multimedia and entertainment such as gaming. In June 2013, it was estimated that Filipinos spend 50 minutes out of an average of 171 minutes daily on entertainment activities such games. [1] This shows that Filipinos spend a big amount of time on smart phones playing games and using applications with similar entertainment value. Furthermore, an application that has game components and offer gaming experience attracts more users and encourages them to use the application on a regular basis.

An application's attractiveness to users is important however it should also serve its purpose. For example, mobile health applications for diabetics should offer entertainment value and at the same time engage users to be more conscious about their health.

Feedback, rewards, and achievements systems embedded in different games have been proven to be effective tools to motivate users perform certain tasks. [14] For example, a computer game, Fish 'n' steps gives feedback and rewards to the user for utilizing walking as a regular physical activity.

Khaled et al. defined persuasive technology as "any interactive product designed to change attitudes or behaviours by making desired outcomes easier to achieve". [15] Persuasive components can be applied to mobile health applications to facilitate and encourage behavioral change. This change in behavior can be used as an instrument to improve user's lifestyle.

2 REVIEW OF RELATED LITERATURE

Among all platforms in 2017, Android, led in mHealth applications. 158,000 health applications in total were available in the Play Store during 2017 which was an increase of 50% from the previous year. This allowed Android to overtake iOS as the leading platform based on the number of mobile health applications. On the other hand, the number of applications available on the App Store increased by 20%. [17]

The increase in the number of health applications in both Play Store and App Store shows that developers are taking interest in mHealth development. More users will be able to use these applications if they are available in leading platforms. Evidently, Android and iOS have the best reach in distributing these mobile health applications. Despite the increase in the number of mHealth applications in Apple's App Store and Google's Play Store, research suggests that these applications are not meeting the needs of patients or clinicians. [8] Listed below are the four (4) major policy issues needed to be addressed by mHealth applications.

(1) Safety - prioritizes user privacy, provides correct diagnosis, does not make false claims
(2) Evidence catalog - can be compared to other applications without considering the ratings given, must have a standardized label and description based on community feedback
(3) Interoperability - data can be stored, saved, and exported
(4) Incentivizing value - application must improve care and value through the use of incentives

2.1 *Persuasive mHealth applications*

One of the simplest ways to alter a person's behavior is through reminding them. This may work for some but others will need more than just alarms to change their behavior. Persuasive technology designers consider the interaction of motivation, ability, and triggers.[18] Triggers come in different forms such as goals set by users and reminders prompted to them. Syncing the application to social media and online forums can also help. Having a community that shares the same goals and contribute feedback also motivates users. Additionally, research suggests that feedback and reward systems can be effective tools for behavior change. [14] With feedback and reward, users can be more involved in the process of changing one's behavior.

This means that applications which have persuasive components in their design can be utilized in health applications in terms of behavior change. Applications can be used to influence users to change their unhealthy lifestyle and be conscious of their health. Success of persuasion in mobile applications can be seen in different applications with the same concept such as CARROT Fit, CARROT To-do, etc.

A study conducted using top-ranked mobile applications that encourage behavior change in 2013 showed that one of the most common techniques used is providing feedback for performance. The applications were also later classified into two (2) types which were educational and motivational. [11] Moreover, according to Singer et al., gameficiation is an effective tool to motivate and influence users in a technology setting. [19] However, it is worth noting that a persuasive application must be used consistently to be effective. This can be a drawback in persuasive applications as the application use time should be considered.

Quittylink, a mobile application developed to help quit smoking was able to promote self-awareness, selfmotivation, and support for people who wanted to quit smoking. The interactive design of the application includes tracking of cigarettes smoked and resisted. Data in the application were also in a graph to help keep track of their progress in terms of smoked and resisted cigarettes. [16]

Altmeyer et al. introduced public display in a mobile application as a means of motivation for users to increase their walking activities. In their work, results show that there was a significant increase in step count of users. A gamified system with the addition of public display was able to encourage these users to walk more and increase their daily step count. [7]

A similar application, HealthyTogether developed by Chen et al. studied the effectiveness of social incentives for lifestyle change for obese and diabetic patients. This application counts the average step count of pairs of users together with the number of floors climbed in a day. One of the dimensions of social incentives employed in the application is an in-game reward system which lets the users win badges based on their average step SugarTraces: A Persuasive technology-enabled mobile application for diabetics UP Diliman, May 2019, Quezon City, Philippines count and floors climbed. Users can also send cheering or taunting messages to others which act as a feedback system and add competitive component to the application. The results show an increase in daily floors climbed which signifies that users tend to walk more. [9]

Table 1 shows the evaluation of diabetes health applications with respect to the policy issues mentioned. Most applications satisfy the four policies however, the design of the basic functionalities are too complicated for users. Meanwhile, in terms of evidence catalog policy, all the applications listed are inconclusive since no published research supporting their claims was found.

Table 1. Evaluation of some diabetes mobile applications with respect to the different health policies.

Application	Safety	Evidence Catalog	Interoperability	Incentivizing Value	Entertainment Factor
Sugar Sense	No	Inconclusive[1]	Yes	Yes	No
Diabetes: M	No	Inconclusive[1]	Yes	Yes	No
Sugar Diary	No	Inconclusive[1]	No	Yes	No
Blood Glucose Tracker	No	Inconclusive[1]	Yes	Yes	No

[1]No published data found

Sugar Sense, an iOS application, satisfies two of the policies stated as seen in Table 1. First, upon reviewing its privacy policy, it was found out that the application collects user data. According to Sugar Sense's privacy policy, anonymous aggregate data reports are collected from its users. This means that user data such as blood glucose, blood pressure, weight, and other health information saved through the application are automatically collected. Meanwhile, interoperability policy is met since the options to store, save, and export their health data are provided. Finally, the application gives health tips and suggestions after an entry is saved and thus satisfies the incentivizing value policy.

Diabetes:M, another diabetes application, also collects data from its users. According to the application's privacy policy, it does not collect any personal information from the device where it is installed without user permission. The application also allows users to export and import data as CSV files which passes the interoperability policy. The application gives a very comprehensive and complete picture of the user's health with the health data that they collect. This satisfies the incentivizing value policy.

Sugar Diary, based on their privacy policy, collects personal data including but are not limited to advertising ID, IP address, and location. The application allows export of data but does not have a way to import data. It is simple enough for people who want to keep track of their blood glucose levels.

Blood Glucose Tracker, an Android application, allows users with subscription to be able to sync its data to multiple devices through Google account. Additionally, it allows export of data through email. It also provides a good amount of information for overall health condition collected from user input.

The diabetes health applications listed above provide the basic functionality of saving and storing blood glucose entries. Entertainment factor, which encourages users to continuously use the application, is missing in the stated applications.

3   SUGARTRACES: A GAMIFIED DIABETES MANAGEMENT APPLICATION

SugarTraces is a mobile health application which aims to help users with prediabetes and diabetes to live a healthier lifestyle by maintaining healthy blood glucose levels. A feedback system is implemented as a response whenever the user inputs a blood glucose reading. Entered readings are recorded and stored locally in the device. Moreover, the feedback system is made up of different statements, which the application use as a response to the user's input. These also make the application more amusing and entertaining to use. The statements vary depending on the level of the entered reading. Achievements can also be unlocked based on the activity of the user. Additionally, the user's profile will also display the streak of consecutive number of days with normal readings. These reward systems are proven to be effective tools in behavior change. [14] The application also displays the most recent entries in a graph.

3.1   *Health policies*

SugarTraces satisfies the three (3) major policy issues namely: safety, interoperability, and incentivizing Value in mHealth applications. The evidence catalog policy will be attempted to be met since a testing phase of the application will be administered.

(1) Safety - no user data collected, entered blood glucose readings are saved only in the device, diagnosis only in terms of tiered blood glucose levels
(2) Interoperability - blood glucose readings are stored and saved in the device, user has the option to share saved blood glucose readings through email
(3) Incentivizing value - provides incentives such as feedback, streak display, and achievements for maintaining healthy blood glucose levels

3.2 Software architecture

The software architecture of SugarTraces consists of only three (3) layers namely: view, logic, and storage. This is illustrated in Figure 1. The view layer is what the user sees on the screen. This includes the five (5) tabs the user can navigate to. This shields the user from the logic layer where the basic functionalities are handled.

Meanwhile, the logic layer manages all the responses needed for every user input. This includes how and when the feedback is chosen and presented, how streaks are computed, and how achievements are unlocked. Each of the game components such as streaks, feedback statements, and achievements has its own logic.

The streaks are based on the consecutive number of days the user entered a normal blood glucose reading. It is terminated if a below normal or above normal reading is entered.

Meanwhile, the feedback displayed on the screen is dependent on the collection of statements divided into three categories: Below Normal, Normal, and Above Normal. A feedback statement will be displayed on the screen depending on the blood glucose reading entered by the user.

Finally, the achievement system is a collectible type where the user can unlock different icons for display depending on their progress with entered readings. The last layer in this architecture is the storage layer. This layer handles data storage and management.

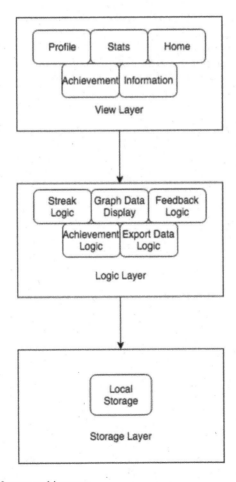

Figure 1.   SugarTraces software architecture.

3.3 Design

SugarTraces consists of a navigation bar with five (5) tabs as shown in Figure 1. These tabs are home, stats, profile, achievements, and information.

The home screen houses the main functionality of the application. In this screen, the user inputs the blood glucose reading. Readings can be classified into three (3) levels: Below Normal, Normal, and Above Normal, as defined in Table 2. As the user inputs a new entry, the application responds with a feedback statement based on the classification of the reading. The flow of the processes involved in data entry is shown in Figure 2.

The feedback statements that the application present to the user are all written by the researchers. These statements are made to encourage the user to consistently use the application and to be entertained at the same time. They are to be evaluated and improved as deemed necessary. These are also the main persuasive component of the application and therefore need to be properly and effectively designed and written.

Additionally, game elements such as streaks and achievements were added to the application. Streaks are incremented every time the user enters a normal blood glucose reading. The date is considered for a streak to be counted. Furthermore, the purpose of achievements and streaks is to motivate the user to continuously use the application and at the same time maintain healthy blood glucose levels. Different achievements can be unlocked based on application usage and data input.

Table 2. Blood glucose levels.

Range	Level
Below 70mg/dL	Below Normal
70 mg/dL to 150 mg/dL	Normal
Above 150mg/dL	Above Normal

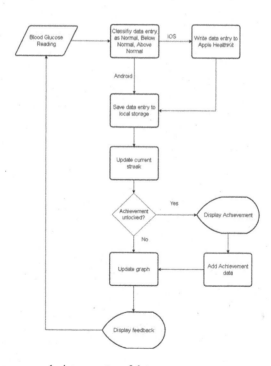

Figure 2. Flow of the processes during an entry of data.

Meanwhile, the profile screen contains information about the total number of entries the user has entered using the application. It also contains information about the user's last recorded reading, number of unlocked achievements, and current streak. The stats page presents the past six (6) readings in a graph. Finally, the achievements screen houses all unlockable badges the users can receive upon specific activity using the application. Initially, there are eight (8) achievements that users can unlock but more will be added in the future. The achievements are listed below.

1. Entered the first normal reading
2. Entered 4 consecutive normal readings after an above normal reading
3. Entered 5 consecutive normal readings
4. Entered 4 consecutive normal readings after a below normal reading
5. Entered 10 consecutive normal readings
6. Had a 3-day normal blood glucose streak
7. Shared data
8. Read the acknowledgement

Users also have the option to share their data via email to others, such as family members, friends, researchers, doctors and medical practitioners.

4   DISCUSSION

Development of the application was done using React Native which allows support for both Android and iOS devices. React Native is chosen for its good cross-platform development.

There are other cross-platform solutions available such as Xamarin, Native Script, and Ionic. Among these, React Native was chosen mainly because of availability of various modules in its large community of users. Some of these cross-platforms solutions are also available as open source projects but some of their services require paid subscription. React Native on the other hand has various libraries available for free.[6]

The application is easier to maintain since up to 70% of the code can be used for both Android and iOS.[6] Additionally, platformspecific code is also easy to write. In a case where a native iOS and React Native version of the same application was developed, performance measured in CPU, GPU, and memory were almost identical.[4] SugarTraces has few functionalities which require less and simpler elements to be rendered. It takes up less than 20MB of storage space when installed on an Android device and around 10MB on an iOS device.

4.1   *HealthKit*

For iOS, the application is integrated with Apple's HealthKit. Apple HealthKit primarily focuses on digital health. They are also partnered with Fitbit and Nike to allow users to share health data with reputed doctors using their applications. It also allows the users to keep track of various health data which provides a complete picture of their health. In line with this, Apple HealthKit is able to collect data from medical applications such as lab results, nutrition data, medication, etc. Overall, HealthKit is able to provide detailed information about the user's health.[3]

SugarTraces uses Apple HealthKit for data storage and export. Only blood glucose, date of birth, and sex are exported from HealthKit's data. This is done by requesting permission from the user after application installation. Blood glucose entries entered in the application are stored locally in the device and in Apple's Healthkit if writing permission is allowed.

4.2 Google fit

On Android, Google Fit, a feature similar to Apple's HealthKit is available. Google Fit focuses more on medical data. It has sensor, record, and history based APIs that help in collecting data using device sensors.

Google Fit lists blood glucose under Health Data Types. This kind of data type in Google Fit is restricted which means that only few applications are allowed to access them. Unfortunately, SugarTraces does not support Google Fit at the moment due to Google currently not allowing new applications for access. This was one of the major problems that the researchers faced in designing the application. This means that the application's data is not integrated through Google Fit for Android users. All data handled in Android are stored locally in the device instead.

4.3 Testing methodology

The objective of testing SugarTraces is to determine its potential to help manage diabetes using gamified components such as streaks and achievements combined with snarky feedback. The following are the key elements prepared for the testing phase:

(1) Questionnaire

A questionnaire was prepared and evaluated using Cronbach's Alpha. Cronbach's Alpha measures internal consistency, relatedness, and reliability of the items in the questionnaire. [10] The items in the questionnaire score the components based on the tester's opinion. It consists of 19 statements to be evaluated on a scale of one (Strongly Disagree) to five (Strongly Agree).

(2) Test Participants

There were a total of forty-five (45) participants with ages ranging from twenty (20) to eighty (80) years old with varying medical conditions related to diabetes. Some these conditions include diagnosed with diabetes, has prediabetes, family history of diabetes, regular blood sugar check up, etc. The distribution of the respondents according to age and medical condition are shown in Table 3 and 4.

(3) Test Execution

(4) The test participants were asked to:

Download and install SugarTraces from Play Store or App Store
Explore and use the application for at least five (5) days
After five days, fill out and submit the prepared questionnaire

4.4 Results and analyses

The fourty-five (45) test participants were clustered and grouped by age and medical condition as seen in Tables 3 and 4. Majority or thirty (30) of the testers are female while fifteen (15) are male. Additionally, twenty-five (25) are Android users while twenty (20) are iOS users.

Table 3. Test participants grouped by age.

Age	Count
20-29	9
30-39	4
40-49	7
50-59	14
60-69	7
70+	4
Total	45

Table 4. Test participants grouped by medical condition.

Age	Diagonosed	Prediabetes	Med. Practitioner	Others
20-29	0	2	0	7
30-39	3	0	0	1
40-49	4	0	1	2
50-59	9	2	1	2
60-69	4	3	0	0
70+	3	1	0	0

[2] Others include: has a family history of diabetes, regularly checks blood sugar

The survey questionnaire focused on the persuasive component of SugarTraces and the usability of the application. These components include the streak, feedback messages, and achievements. The most important statements evaluated by the testers are the following:

1. I found the feedback messages appealing.
2. I was motivated by the positive feedback messages to maintain healthy blood sugar level.
3. I was motivated to keep a streak going after starting one.
4. It was challenging to complete the achievements.
5. I found application helpful in managing diabetes.
6. I would recommend the application to other diabetics.

The average scores of the statements mentioned above are shown in Table 5.

Based on the average scores computed from the questionnaires, the mobile application, SugarTraces scored well when it comes to the specific persuasive components being evaluated. The feedback messages displayed after entering a blood sugar reading were revealed to be appealing to the testers. The positive feedback messages also helped in motivating users to maintain healthy blood sugar level. Users are also motivated in keeping a streak to maintain healthy blood sugar levels when using the application. However, this score is relatively low compared to others because some of test participants initially did not know what the streak was for despite having an animation indicator to explain what it is. Meanwhile, the achievements were found to be challenging to complete but they also recommended to add more achievements to be completed through the application. Overall, they found the application helpful in managing diabetes and would recommend it to other diabetics.

Another important issue that was evaluated in the testing was whether users needed technical assistance in navigating through the application. It is important to note that aside from the installation instructions provided, no further assistance was given to the testers in navigating through the application.

The average score for the statement "I did not need technical assistance in navigating through the application." was computed to be 3.82. This score shows that it is easy to navigate

Table 5. Average Scores.

Statement	Score
I found the feedback messages appealing.	4.31
I was motivated by the positive feedback messages to maintain healthy blood sugar level	4.24
I was motivated to keep a steak going after one.	3.91
It was challenging to complete the achievements.	4.33
I found application helpful in managing diabetes.	4.64
I would recommend the application to other diabetics.	4.62

[3] 1 – lowest score, 5- highest score

through the application without technical assistance. Howver, again, this is low relative to others due to the scores given by the testers in the 70 and above age group. This shows that people in the said age group require more technical assistance compared to others.

Looking into more detail of the scores computed, Figure 3 shows a graphical representation of the average scores clustered into age groups. The graph shows that the users who fall to the age group of 70 and above have given the lowest scores. In addition to this, the same group of users also "felt uncomfortable in using the application"giving the statement an average score of 3.0. This score is low because the users find it hard in general to navigate through any mobile application.

The average scores of two(2) statements that are related to the overall user experience are compared across platforms (Android and iOS). These statements are the following:

1. I did not need technical assistance in using the application.
2. I found it easy to navigate through the application.

Looking at Figure 4, iOS users scored higher for both statements. The iOS users found it easier to navigate through the application and are more confident in using the application without technical assistance than the Android users.

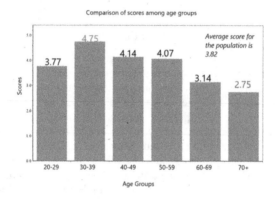

Figure 3.   Comparison of average score among age groups
[4] 1 – lowest score, 5 – highest score.

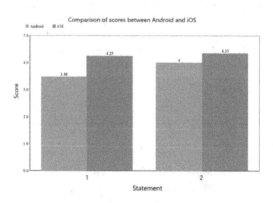

Figure 4.   Comparison of Average Score Between Android and iOS
[5] 1 – lowest score, 5 – highest score.

5 CONCLUSION AND FUTURE WORK

SugarTraces shows that feedback messages as a component of behavioral change techniques in diabetes applications help motivate users to maintain healthy blood sugar level. Additionally, appealing feedback messages that give the users the interactive feel to the application also help in promoting behavior change through mobile health applications.

Gamification elements such as streaks and achievements assist in making the application stick to the users. These elements make the users want to use the application more and can be applied to mobile health applications that need regular user activity.

However some of the components of the application can be improved based on the feedback of the participants after the testing phase. First, to address the streak being counterintuitive, it is suggested to make UI changes such as making the icon animations flashier and easy to be noticed. Second, the low score in terms of technical assistance can be addressed by revamping the on boarding screens to include more information in navigating through the application. Finally, more achievements will be added for users to complete and make them use the application more.

To conclude, SugarTraces helps manage diabetes through the integrated persuasive and gamified components. It has the potential to stick to the user for continuous usage in comparison to the current typical and traditional blood sugar management applications in the market with basic blood glucose monitoring and no entertainment factor.

REFERENCES

[1] 2013. How much time do Filipinos spend on their smartphones? Retrieved January 23, 2019 from https://www.rappler.com/business/39027-filipinossmartphones-nielsen-report

[2] 2016. Share of the population that uses a mobile phone in the Philippines from 2014 to 2020. Retrieved January 2, 2019 from https://www.statista.com/statistics/570389/philippines-mobile-phone-user-penetration/

[3] 2018. Apple's HealthKit vs Google Fit: How do they compare against each other? Retrieved December 21, 2018 from https://medium.com/@promatics/appleshealthkit-vs-google-fit-how-do-they-compare-against-each-other8deaa709b540

[4] 2018. Comparing the Performance between Native iOS (Swift) and React-Native. Retrieved December 28,2018 from https://medium.com/the-react-nativelog/comparing-the-performance-between-native-ios-swift-and-react-native7b5490d363e2

[5] 2018. Deaths in the Philippines, 2016. Retrieved December 28,2018 from https://psa.gov.ph/content/deaths-philippines–2016

[6] 2018. Xamarin vs React Native vs Ionic vs NativeScript: Cross-platform Mobile Frameworks Comparison. Retrieved January 6, 2019 from https://www.altexsoft.com/blog/engineering/xamarin-vs-react-native-vsionic-vs-nativescript-cross-platform-mobile-frameworks-comparison/

[7] Maximilian Altmeyer, Pascal Lessel, Tobias Sander, and Antonio Krüger. 2018. Extending a Gamified Mobile App with a Public Display to Encourage Walking. (2018).

[8] David W. Bates, Adam Landman, and David M. Levine. 2018. Health Apps and Health Policy What Is Needed? JAMA (Oct. 2018). https://doi.org/10.1001/jama. 2018.14378

[9] Yu Chen, Mirana Randriambelonoro, Antoine Geissbuhler, and Pearl Pu. 2016. Social Incentives in Pervasive Fitness Apps for Obese and Diabetic Patients. (2016).

[10] Cronbach's Alpha [n. d.]. What does Cronbach's Alpha Mean? Retrieved May 12, 2019 from https://stats.idre.ucla.edu/spss/faq/what-does-cronbachs-alpha-mean/

[11] Jaclyn Maher David Conroy, Chih-Hsiang Yang. 2014. Behavior Change Techniques in Top-Ranked Mobile Apps for Physical Activity. American Journal of Preventive Medicine (2014).

[12] Department of Health [n. d.]. DOH Leads World Diabetes Observance in the Philippines. Retrieved December 21, 2018 from https://www.doh.gov.ph/node/11786

[13] Department of Health [n. d.]. What are the leading causes of mortality in the Philippines? Retrieved December 10, 2018 from https://www.doh.gov.ph/node/1058

[14] Feng Gao. 2012. Design for Reflection on Health Behavioral Change. In Proceedings of the 2012 ACM international conference on Intelligent User Interfaces. ACM New York, NY, USA, 379–382.

[15] Rilla Khaled, Robert Biddle, James Noble, and Pippin Bar. 2006. Persuasive Interaction for Collectivist Cultures. In AUIC '06 Proceedings of the 7th Australasian User interface conference, Vol. 50. Australian Computer Society, Inc. Darlinghurst Australia, Hobart, Australia, 73–80.

[16] Jeni Paay, Jesper Kjeldskov, Mikael B. Skov, Nirojan Srikandarajah, and Umachanger Brinthaparan. 2015. Quittylink: Using Smartphones for Personal Counseling to Help People Quit moking. (2015).

[17] Research 2 Guidance 2017. 325,000 mobile health apps available in 2017 âĂŞ Android now the leading mHealth platform. Retrieved December 19, 2018 from https://research2guidance.com/325000-mobile-health-apps-available-in-2017/

[18] Monica Rozenfeld. 2018. How Persuasive Technology Can Change Your Habits. Retrieved January 6, 2019 from http://theinstitute.ieee.org/technology-topics/consumer-electronics/how-persuasive-technology-can-change-your-habits

[19] Leif Singer and Kurt Schneider. 2012. It Was a Bit of a Race: Gamification of Version Control. In Proceedings of the 2nd international workshop on Games and software engineering (2012). Retrieved December 18, 2018 from https://www.researchgate.net/publication/230854785_It_was_a_ bit_of_a_ race_Gamification_of_version_control

[20] Nancy Staggers and Mattias Georgsson. 2015. An Evaluation of patients'experienced usability of a diabetes mHealth system using a multi approach. Journal of Biomedical Informatics 59 (Nov. 2015). https://doi.org/10.1016/j.jbi. 2015.11.008

[21] World Health Organization 2018. Diabetes. Retrieved December 10, 2018 from https://www.who. int/news-room/fact-sheets/detail/diabetes

Theory and Practice of Computation – Nishizaki et al (eds)
© *2021 Taylor & Francis Group, London, ISBN 978-0-367-41473-3*

Telecollaboration to promote preservation of Asian local indigenous knowledge and intercultural communicative competence

A.P. Vilbar, R. Yaakub, N. Ahmad, R. Makaramani, Z. Bayarchimeg, M. Laus,
C. Malaque & J.E. Gumalal
University of the Philippines Cebu, Cebu City, Philippines

ABSTRACT: Local Indigenous Knowledge Systems (LINKS) are local body of knowledge practiced by people with their natural environment which are essential to modern living. However, LINKS have deteriorated due to dominance of Western science and prejudice. This collaborative research among 144 students and professors from Malaysia, Mongolia, the Philippines and Thailand examined the affordances of telecollaboration in preserving the four countries' LINKS by producing an e-book. It also determined the effect of telecollaboration on students' Intercultural Communicative Competence (ICC). First, the students conducted a LINKS inventory in their countries. Then, they analyzed the dis/similarities of the four countries' LINKS through telecollaboration tools. Data were collected from videoconferencing evaluation, Facebook group posts, post-project questionnaire, and interviews. Findings show the four countries have similar LINKS on health and pregnancy, navigation, parenting style, farming/fishing and risk-disaster. The use of telecollaboration tools depends upon the nature of the tasks. Social media and emails are used for critical reading and instant feedbacking while videoconferencing is used for wider dissemination and reflections. The project developed the participants' ICC with specific elements on openness to learn other cultures; knowledge of their own and other social groups' practices; and critical cultural awareness to evaluate in one's own and other cultures. Their active telecollaboration widened their picture of understanding their own and other cultures in the context of global understanding and mutual respect.

1 INTRODUCTION

In 2004, in Sumatra, Indonesia, an earthquake caused a tsunami that resulted in nearly 300,000 deaths. In effect, the Moken community in the Surin Islands located in between Myanmar and Thailand, was devastated however none of the community was lost. The Moken claimed that were saved by the legend of the god of waves Laboon who was believed to visit once in every two generations. Using this legend and by observing the changes in the ocean, the community proceeded to higher ground before the tsunami struck (Stevens 2009).

This Laboon legend which became the tsunami warning is considered as local indigenous knowledge systems (LINKS). LINKS are body of knowledge and traditional technologies, possessed by people with long histories of interaction with their natural environment. Like the Laboon tale, LINKS are intimately tied to spirituality (Mexico City Synthesis Workshop "Science and technology for sustainable development". 2002, 2002) and are associated with environmental management and cultural norms (Dei 2011). They include present technologies that can guide the present generation (Zinyeka 2013).

However, these knowledge are perceived to be of little value (Khumalo & Baloyi 2017) and have deteriorated due to content assimilation which resulted from the loss of interest from the young generation (Donato-Kinomis 2016). LINKS are not strongly integrated in local education due to the dominance of Western science over other ways of knowledge (Khupe 2014).

Addressing the need to preserve LINKS through education, the researchers conducted a telecollaboration among the four countries in Malaysia, Mongolia, the Philippines and Thailand to create an electronic book (e-book) that featured the common LINKS among the four countries. Telecollaboration is the use of computer and/or digital communication tools to promote learning through social interaction and collaboration (Corbett 2003). The researchers are recipients of the workshops on promoting telecollaboration and Education for sustainable development conducted by UNESCO Thailand in 2013 and 2014. They are professors from Universiti Sains Malaysia in Malaysia, Suan Sunandha Rajabhat University in Thailand, Mongolia State University of Education in Mongolia, and the University of the Philippines High School Cebu in the Philippines. These professors were trained on the integration of LINKS in the curriculum and conducting telecollaboration.

This study aims to analyze the experiences of the students and professors among four countries in conducting the telecollaboration project. It seeks to develop a Project-Based Learning Telecollaboration Model which aims to produce an ebook as an output that documented the common LINKS among the four countries. It also aims to determine the effect of the telecollaboration on the students' intercultural communicative competence (ICC). ICC is the ability to communicate effectively and appropriately among people in a culturally diverse context and the ability to mediate between cultures enabling students to reflect on themselves from an external perspective while analyzing their own values and beliefs (Corbett 2003).

2 RESEARCH QUESTIONS

This research developed a Project-Based Learning Telecollaboration (PBLT) Model which served as a platform for online telecollaboration among Malaysian, Mongolian, Filipino, and Thai students in creating an ebook that featured the similar LINKS among the four countries. It also examined the kinds and nature of telecollaboration tools and determined the effect of the telecollaboration project on the students' intercultural communicative competence.

3 LITERATURE REVIEW

Telecollaboration is the use of different online communication tools such as video conferencing, social networking sites, emails that allow participants to collaborate and work, benefiting from the virtual dialogs (Andujar 2019).

This research adapts intercultural model of telecollaboration of O' Dowd (2016) that uses telecollaboration to promote intercultural learning by comparing and analyzing parallel local cultural texts among four international partner institutions. This model can provide authentic and effective way to experience cross-cultural interactions to develop intercultural learning which is the demand of the increasingly interconnected world (Hsu & Beasley 2019).

3.1 Project-based learning

Telecollaboration anchored on PBL can lead to optimum results for PBL is a systematic teaching method that engages students in learning knowledge and skills through an inquiry process structured around complex, authentic questions and carefully designed products and tasks (William et al. 1999). PBL uses authentic, complex, and real-life tasks to motivate learning and provide learning experiences These real-life tasks include comparison and analysis tasks in which students analyze cultural products of both cultures, and the collaborative tasks in which students create together a final product (R. & P., 2009). Telecollaborators maximize the affordances of technology by using email, social networking sites, and video-conferencing (Develotte, Guichon, & Vincent 2010).

The output of this project is an electronic book that features the common LINKS among the four countries. Ebooks are electronic versions of academic books, and other one-off publications which are lighter and more portable than regular textbooks (Enright 2014).

3.2 *Byram's intercultural communicative competence*

To determine the impact of telecollaboration on the students' ICC, this study uses the Byram's ICC model which is the ability to interact and communicate effectively and appropriately in people in a culturally diverse context; and the ability to mediate between cultures enabling students to reflect on themselves from an external perspective while analyzing their own values and beliefs (Byram 1993). ICC trains learners to be 'diplomats', able to view different cultures from a perspective of informed understanding" and involves interrelated five elements (Corbett 2003):

- Attitudes: curiosity and openness, readiness to suspend disbelief aboutother cultures and belief about one's own;
- Knowledge: of social groups and their products and practices inone's own and in one's interlocutor's country;
- Skills of interpreting and relating ability to interpret a document or eventfrom another culture, to explain it and relate it to documents from one's own;
- Skills of discovery and interaction: ability to acquire new knowledgeof culture and cultural practices and the ability to operate knowledge,attitudes and skills under the constraints of real-time communicationand interaction;
- Critical cultural awareness: an ability to evaluate critically and on thebasis of explicit criteria perspectives and practices and in one's ownand other cultures and countries

4 THE TELECOLLABORATION PROJECT

There were 136 students and 8 professors involved in the project: Universiti Sains Malaysia (USM) had 8 students and 2 professors; Mongolia State University of Education (MSUE) had 11 students and 1 professor; Suan Sunandha Rajabhat University (SSRU) in Thailand had 78 students 1 professor; University of the Philippines Cebu High School (UP Cebu) had 39 students and 4 professors. The professors are recipients of the UNESCO Thailand's Telecollaboration Workshop on Reorienting Teacher Education towards Education for All and Education for Sustainable Development.

This telecollaboration projected was integrated in each country's course for one semester. For SSRU, it was in Educational Technology; for MSUE, it was in Teacher Education; and for USM, it was in Malaysian Culture and in Language Teaching. For the Philippines, it became a multidisciplinary project in English, Research, Social Sciences, and Science subjects. The students were grouped into the five common LINKS: 1 (Risk-Reduction Practices), 2 (Pregnancy and Health), 3 (Navigation), 4 (Farming), and Group 5 (Fishing). Each group was comprised of Malaysians, Mongolians, Thai, and Filipinos. There were three telecollaboration sessions conducted in ten months.

5 METHODOLOGY

This descriptive research used qualitative method. To develop the Project-Based Learning Telecollaboration Model, we analyzed the experiences and evaluation data of the professors and students of the three telecollaboration sessions. To determine the effect of the project on students' ICC, we analyzed the students' 136 responses of the open-ended questionnaires and 40 Facebook Group posts using Byram's ICC framework. This research also used the Participant observation under Participatory Action Research. In Participant observation, the researchers become part of the research process being observed and immersed in the setting of the social situation with the student-participants (Marshall, Permalink, & Lynch 2011); thus, the experiences and assessment of the professors and researchers are essential to the data.

The Framework Analysis procedure of (Ritchie & Spencer, 1994 as cited in Srivastava & Thomson, 2009) was adapted in analyzing the responses and posts: (1) Familiarization, (2) Identifying the theme, (3) Indexing, and (4) Interpretation. For research ethics, the names and pictures of the students in the Facebook group posts were held confidential; and the responses of the open-ended questionnaires did not reflect the names of the students.

6 RESULTS AND DISCUSSION

6.1 *The project-based telecollaboration model to develop the ebook*

The Project-Based Telecollaboration Model of Vilbar, et. al (2019) in developing and publishing the ebook followed this process as shown in Figure 1: (1) Conduct LINKS inventory in every country, (2) Discuss through telecollaboration tools social media and emails, (3) Conduct videoconferencing sessions, (4) Design and publish the eBook.

In Stage 1, the professors from four different countries integrated the ebook project in their respective syllabi. In this stage, the students conducted LINKS inventory through conversational methods to their specific countries and submitted the analyses of their LINKS to their professors.

In Stage 2, the telecollaboration among the four countries commenced. The students were grouped into five. Each group had Malaysians, Mongolians, Thais and Filipinos. The groups were the Risk-Reduction Practices, Pregnancy and Health, Navigation, Farming, and Fishing. The groups used social media Facebook Group and emails as telecollaboration tools in achieving their tasks to compare and contrast the four LINKS and analyzing the LINKS' social, cultural, and historical perspectives. On the other hand, the groups used emails for sending summaries of LINKS or large files like the draft of the ebook.

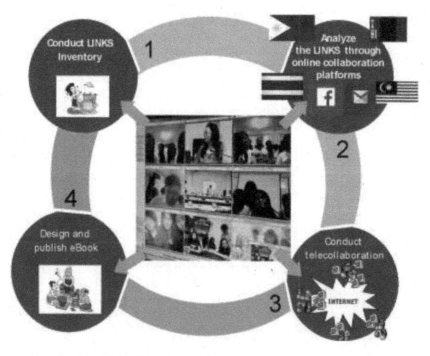

Figure 1.   The project-based telecollaboration model in developing intercultural eBook.

In Stage 3, the students and professors conducted three videoconferencing sessions. The videoconferencing platform used was Aculearn, provided by Suan Sunandha Rajabhat University, Thailand. The videoconferencing served as the online platform to present the group summary of the LINKS to the whole groups, to discuss their analysis of the LINKS, and to share their reflections about the LINKS. There were three videoconferencing conducted which lasted for 3-4 hours per session.

The stages show that the affordances of the telecollaboration tools depend upon the task objectives in each stage. For instance, in Stages 1 and 2, the researchers used Facebook group and emails due to the nature of the collaboration tasks which required instant and spontaneous social communication and critical analysis of the four countries' LINKS. Considering that these students have different nationalities and were strangers in the start of the project, there was a need to use Facebook as a social media platform for them to communicate and accomplish the tasks in a telecollaboration platform that promotes lesser anxiety and instant communication (Halpern & Gibbs 2013). Like a virtual window to global socialization, the socio-cognitive nature of Facebook Group Post allowed the students to develop personal relationships among the four nationalities thus allowing ease of communication to accomplish the tasks (Lambic 2016; Ellison, Steinfield & Lampe 2007; Dyson, et al. 2014) as shown in the post in Figure 2.

In this figure, some Malaysian, Thai, and Mongolian students made a spontaneous conversation comparing their ages and university courses. As icebreakers in the communication thread, GIFs were used to promote self-identities, humor, and visual metaphors among the thread community (Nooney, Portwood-Stacer, & Eppink 2014). GIF refers to the Graphics Interchange Format which is typically used to mean an animated GIF file or an otherwise short, silent, looping, untitled moving image.

Furthermore, in this stage, critical analysis was better done in reading Facebook posts and messages and emails than in analyzing the LINKS through videoconferencing. Compared to videoconferencing, FB and emails can allocate time to read the messages and reply to the thread or email based on their understanding of the LINKS (Monk & Watts 1998). On the other hand, the emails were used when the students sent large files of their collected LINKS. The nature of files included a summary of LINKS in each country, comparison of LINKS among the four countries, and the draft designs of the ebook.

In addition, the four countries speak varieties of English and LINKS use many indigenous languages which are difficult to spell and pronounce from the foreigners' perspectives. Writing messages in the Facebook group or chat support can avoid the potential miscommunication problem of phonological variations. This assertion was validated by Mongolian student Khulan when she said that the telecollaboration expanded her knowledge on cultures and that the FB comments motivated her to improve her English.

In Stage 3, videoconferencing became the preferred medium due to the nature of the task which was to present the group summary of LINKS to the whole groups. The Aculearn videoconferencing has the feature that can enable the students and professors to view the documents or presentation slides while the groups are presenting. It allowed wider understanding of the LINKS presentations because students can see the presenters, analyze the data, and can received instant feedback with the presenters simultaneously (Denstadli, Julsrud, & Hjorthol 2012).

The videoconferencing became a culminating venue to share the students' tasks in completing their project. Group presentations were comprehensive that promoted excitement and positive atmosphere among the four countries. According to Filipino Student Cleofe, "I'm excited to learn to see what other groups did like the LINKS on pregnancy and calamity. I also want to see how they look like in person."

Mongolian Student Bilguin shared,

"We agree that face to face meeting, Facebook were enjoyable. We heard how Malaysians, Thai, and Filipino students talked in English. We were motivated to work hard on our English."

In Stage 4, selected students from four countries collaborated and designed and published the ebook through emails and Facebook.

Figure 2.   Fb group post conversation among Malaysian and Thai students.

6.2   *The ebook and the common LINKS*

From the telecollaboration, the four countries have common LINKS related to health and pregnancy, parenting navigation, farming, and disaster. Due to its geography, Mongolia has limited LINKS on fishing unlike the three countries that have fishing and coastal communities.

As an example of the many common LINKS, Figure 3 shows that pregnant women among the four countries must eat fruit and vegetables to produce quality breast milk. For Malaysians, pregnant women must have soup with vegetables. For the Philippines, they must eat fruits and vegetables. For Mongolia, they must have soup with boiled sheep bone and drink milk tea with millet. For Thailand, they must have banana blossom soup.

The Philippines and Malaysia have common LINKS of letting pregnant women eat fruits and vegetables which vary depending upon the geographical location of these women. On the other hand, Mongolia and Thailand have specific food to eat. In addition, parents from Malaysia, Mongolia, the Philippines, and Thailand use the stories of spirits or supernatural as a method to discipline their children in not staying at home late.

In the videoconferencing, Filipino students explained, "because of fear of ghosts, children rush home before dusk. In effect, the parents feel that their children are safe." The findings support

Figure 3. Screen shot of the ebook comparing indigenous practices on pregnancy.

the contention that LINKS are intimately tied to social relations and spirituality (Mexico City Synthesis Workshop "Science and technology for sustainable development". 2002, 2002).

6.3   Telecollaboration project on students' intercultural communicative competence

The analysis of three telecollaborations, 136 evaluation with open-ended answers and 40 Facebook Group posts revealed the project promoted intercultural communicative competence: (1) curiosity and openness to learn one's own and other cultures, (2) knowledge of their own and other social groups' practices, (3) critical cultural awareness to evaluate in one's own and other cultures as shown in Table 1.

Curiosity and openness to learn other cultures
The FB Group post interactions in Figure 4 between Malaysian student Silvana and Filipino student Trisha (names of the students are only pseudonyms) reveal their openness to learn each other's cultures. Both ladies discussed the similar indigenous first aid practices in Malaysia and the Philippines.

From the evaluation, Thai students expressed their curiosity about many indigenous practices among the three countries. Paneta said,

"The Philippines surprised me for some stories are the same with Thailand. Malaysia and Mongolia have many interesting cultures."

Anthara shared,

"I learned geographic and global issues of the other countries better." Vitabol learned that some LINKS are strange and funny but are always useful in their context."

The Facebook conversations reveal their readiness to suspend disbelief about other cultures and belief about one's own and willingness not to assume own beliefs are the only possible and correct ones (Byram 1993). Malaysian Silvana and Filipina Trisha both realized that each country has a unique way of doing first aid treatment to burnt skin. The conversations did not

Table 1. Sample reflections of the students.

Intercultural Communicative Competence	Sample Reflections
curiosity and openness to learn one's own and other cultures	"I learned geographic and global issues of the other countries better."
	"I learned that some LINKS are strange and funny but are always useful in their context."
	"I think we have a little similarity about swallowing wet bread in order to remove the bones stuck in the throat. But instead of bread (for Malaysians), we swallow banana in the Philippines."
knowledge of their own and other social groups' practices	"My three classmates knew Malaysia as a country that offers English courses and has products like shampoo, soap, and baby diapers. After the telecollaboration, we learned about the three countries' location, language, and population."
	"At first I was like in a culture shock. But knowing a new and different culture is always interesting. When we shared our topics together through video conferencing, it was fun to know about other countries."
critical cultural awareness to evaluate in one's own and other cultures	"Rice is an imported product in Mongolia. Thais believe that the goddess of grains or Mae Posop takes care of the rice fields. Rice is Thailand's main dish and farmers work very hard to produce it. Thus, elders teach the indigenous practice that if you leave rice on your plate, you will have dark spots on your face. It is a good practice because children want to have a lovely face."
	"The project gave me good tactics when we go abroad".

yield into judgment on which treatment is medically proven, the butter or toothpaste. Rather they focus on their distinctiveness and further discussed the bio-chemical components of the butter, toothpaste, and packed ice for skin healing.

Knowledge of their own and other social groups' practices

The Facebook post conversations between Filipina student Maharlika and Malaysian student Fatima revealed their knowledge of their own and other countries' indigenous health practices as shown in Figure 5. Maharlika initiated the conversation with the idea that both Malaysia and the Philippines have similar practice of swallowing bread or banana to remove the fish bones stuck in the throat. In the second conversation, Fatima was curious about using flying lizards used by Filipino locals to cure asthma. Fatima replied by posting a picture of the lizard locally called, 'hambuhukag'.

The Facebook posts and evaluation demonstrate the students' willingness to learn new cultures in the context of equality and intercultural understanding (Byram 1993). Malaysian Fatima and Filipina Maria expressed their amazement of using banana and wet bread as a first aid when fish bones are stuck in the throat and further discussed on the common LINKS among the four countries. Their openness prove that project developed their' curiosity to other cultures and awareness of other culture's social interactions (Corbett 2003; Byram 1993).

The evaluation revealed that the students developed their knowledge of social groups and their products and practices and in one's country [31]. Thai student Unchalee said, "At first I was like in a culture shock. But knowing a new and different culture is always interesting. When we shared our topics together through video conferencing, it was fun to know about other countries."

Tong shared that in Thailand, there are a lot of beliefs to teach children proper discipline such as avoiding playing Hide and Seek at night for ghosts might take the children away. Achara added that locals have different beliefs. It is good to learn and exchange indigenous knowledge to others.

Figure 4.   FB post between Filipina and Malaysian students discussing health practices.

On the other hand, Mongolian student Ganbaatar said, "Mongolia has many hypocenter earthquakes and Mongolians have the skill to observe the behaviors of animals because they are very sensitive to their environment."

Another Mongolian student Arban shared that before joining the e-book project, seven of her classmates did not have any knowledge about Malaysia, Thailand, the Philippines. She said,

> "My three classmates knew Malaysia as a country that offers English courses and has products like shampoo, soap, and baby diapers. After the telecollaboration, we learned about the three countries' location, language, and population."

The findings reveal that the research promoted ICC which allowed students to mediate between cultures enabling students to reflect on themselves from an external perspective while analyzing their own values and beliefs (Ryshina-Pankova 2018) (Corbett, n.d.). The telecollaboration allowed the students to experience intercultural learning process of widening the picture of their own cultural identity and increasing their knowledge of foreign behaviors and cultures (Corbett 2003; Byram 1993).

Prior to the telecollaboration, the Mongolians never had a background about Filipinos and only equated Malaysia with its soap and shampoo products. However, after the telecollaboration, the Mongolian students were in surprise that the four countries shared some similar LINKS on pregnancy and health. In addition, Filipino students expressed that the research promoted wider understanding about Mongolians and their cultures since the said country is understudied in the Philippine curriculum. The evaluation said that students' own indigenous practices have similarities to other countries and if there are variations, these can be due to their religion, geographical location, and environment.

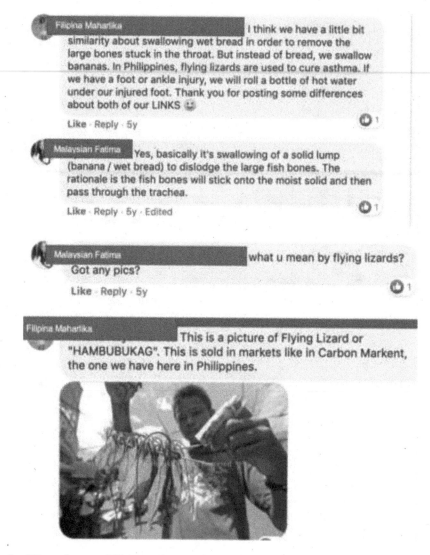

Figure 5.   FB post between Filipina and Malaysian students discussing indigenous practices.

Critical cultural awareness to evaluate in one's own and other cultures

From the evaluation, Thai student Khon said that the project made him understand Asian cultures more. He thought that the indigenous practice of giving warning not to sing in the kitchen to avoid getting married to older persons was unique only among Thais. However, during the telecollaboration, he realized that this practice is common among the four countries whose lesson is to teach table manners. Mongolian student Gerel said that I learned many indigenous practices on producing and eating rice. She wrote,

"Rice is an imported product in Mongolia. Thais believe that the goddess of grains or Mae Posop takes care of the rice fields. Rice is Thailand's main dish and farmers work very hard to produce it. Thus, elders teach the indigenous practice that if you leave rice on your plate, you will have dark spots on your face. It is a good practice because children want to have a lovely face."

Another Thai student Tong reflected, "The project gave me good tactics when we go abroad". This means that he can avoid culture shock if he goes to these countries in the future.

The comments of Tong show that he developed substantial self-reflection on how to negotiate meaningful resolution to avoid potential intercultural conflicts if ever he visits the those three countries (Lambert 1999). The evaluation proved that the research promoted students' critical cultural awareness to evaluate in one's own and other cultures. The students demonstrated their ability to evaluate critically their own and other cultures on the basis of explicit criteria, perspectives, practices and products (Corbett 2003; Byram 1993). This evaluation was in the context of mutual respect of diversity and global understanding.

7 CONCLUSION AND RECOMMENDATION

The research proves that the Project-Based Telecollaboration Model can accomplish the project of producing an ebook that features the common LINKS among Malaysia, Mongolia, Thailand, and the Philippines. This means that the use of ICT resources like telecollaboration coupled with the appropriate educational theory can promote successful educational output.

The research shows that the use of telecollaboration tools depends upon the nature of tasks. Social media and emails are used in collaboration to promote lesser anxiety, instant messaging and feedbacking which leads to spontaneous discussion. Considering that the respondents are adolescents, the use of social media is more appropriate to these digital natives. On the other hand, the use of videoconferencing is used a learning platform to promote wider dissemination of outputs as a culmination of a successful task completion. These modalities allowed the researchers to experience a new realm of collaborative inquiry and re-construction of LINKS.

Furthermore, the research promoted students' intercultural communicative competence with specific elements on openness to learn other cultures; knowledge of their own and other social groups' practices; and critical cultural awareness to evaluate in one's own and other cultures. Their active telecollaboration widened their picture of understanding their own and other cultures in the context of global understanding and mutual respect. This study recommends conducting more telecollaborative projects among other countries and conduct further investigation explaining the socio-cultural background on the given LINKS.

ACKNOWLEDGMENTS

The researchers acknowledged the following institutions, mentors, and students in completing this research work: Universiti Sains Malaysia, Penang, Malaysia; Suan Sunandha Rajabhat University, Bangkok, Thailand; Mongolia State University of Education, Ulaanbaatar, Mongolia; University of the Philippines Cebu High School, Cebu City, Philippines, UNESCO Bangkok Telecollaboration Trainers: Mr. Avelino Mejia Jr., Dr. Jonghwi Park, Mr. Hartfried Schmid, Ms. Lay Cheng Tan, and Ms. Mel Tan.

REFERENCES

Andujar, A. (2019). *Language Learner Engagement in Telecollaboration Environments* (pp. 249–266). https://doi.org/10.4018/978-1-7998-0119-1.ch014

Byram, M. (1993). Introduction. *Language, Culture and Curriculum, 6*(1), 1–3. https://doi.org/10.1080/07908319309525130

Corbett, J. (n.d.). *No Title.*

Corbett, J. (2003). *An Intercultural Approach to English Language Teaching.* Clevedon: Multilingual Matters.

Dei, G. J. S. (2011). Indigenous Philosophies and Critical Education: A READER. *Counterpoints, 379*, 1–13.

Denstadli, J. M., Julsrud, T. E., & Hjorthol, R. J. (2012). Videoconferencing as a mode of communication: A comparative study of the use of Videoconferencing and face-to-face meetings. *Journal of Business and Technical Communication, 26*(1), 65–91. https://doi.org/10.1177/1050651911421125

Develotte, C., Guichon, N., & Vincent, C. (2010). The use of the webcam for teaching a foreign language in a desktop videoconferencing environment. *ReCALL, 22*(3), 293–312. https://doi.org/10.1017/S0958344010000170

Donato-Kinomis, X. G. (2016). Indigenous Knowledge Systems and Practices (IKSPs) in the Teaching of Science. *13th National Convention on Statistics (NCS)*, 1–8.

Dyson, B., Vickers, K., Turtle, J., & Cowan, S. (2015). Evaluating the use of Facebook to increase student engagement and understanding in lecture-based classes. Higher Education, 69, 303e313.

Ellison, N. B., Steinfield, C., & Lampe, C. (2007). The benefits of Facebook "friends:" Social capital and college students' use of online social network sites. Journal of Computer-Mediated Communication, 12(4), 1143–1168.

Enright, S. (2014). Ebooks: the learning and teaching perspective. *Ebooks in Education: Realising the Vision*, 21–33. https://doi.org/10.5334/bal.d

Halpern, D., & Gibbs, J. (2013). Social media as a catalyst for online deliberation? Exploring the affordances of Facebook and YouTube for political expression. *Computers in Human Behavior, 29*(3), 1159–1168. https://doi.org/10.1016/j.chb.2012.10.008

Hsu, S. Y., & Beasley, R. E. (2019). The effects of international email and Skype interactions on computer-mediated communication perceptions and attitudes and intercultural competence in Taiwanese students. *Australasian Journal of Educational Technology, 35*(1), 149–162. https://doi.org/10.14742/ajet.4209

Khumalo, N. B., & Baloyi, C. (2017). African Indigenous Knowledge: An Underutilised and Neglected Resource for Development. *Library Philosophy and Practice (e-Journal)*. Retrieved from https://digitalcommons.unl.edu/libphilprac/1663

Khupe, C. (2014). *Indigenous Traditional Knowledge and School Science: Possibilities for Integration.* (March), 1–251.

Lambert, R. D. (1999). Language and Intercultural Competence. In *Striving for the third Place: Intercultural Competence through Language Education*. Retrieved from http://files.eric.ed.gov/fulltext/ED432918.pdf#page=73

Lambic, D. (2016). Correlation between Facebook use for educational purposes and academic performance of students. Computers in Human Behavior, 61, 313–320.

Marshall, C., Permalink, G. B. R., & Lynch, B. K. (2011). *UCLA Issues in Applied Linguistics Title Designing Qualitative Research by Publication Date*. Retrieved from https://escholarship.org/uc/item/3m25g8j8

Mexico City Synthesis Workshop "Science and technology for sustainable development". 2002, I. S. on S. for S. D. N. 9. (2002). *ICSU Series on Science for Sustainable Development No. 9.*

Nooney, L., Portwood-Stacer, L., & Eppink, J. (2014). A brief history of the GIF (so far). *Journal of Visual Culture, 13*(3), 298–306. https://doi.org/10.1177/1470412914553365

O'Dowd, R. (2016). Emerging Trends and New Directions in Telecollaborative Learning. *CALICO Journal, 0*(0), 291–311. https://doi.org/10.1558/cj.v33i3.30747

R., O., & P., W. (2009). Critical issues in telecollaborative task design. *Computer Assisted Language Learning*, Vol. 22, pp. 173–188.

Ryshina-Pankova, M. (2018). Discourse moves and intercultural communicative competence in telecollaborative chats. *Language Learning and Technology, 22*(1), 218–239.

Srivastava, A., & Thomson, S. B. (2009). Framework Analysis: A Qualitative Methodology for. *Applied Policy Research. JOAAG, 4*(2), 72–79. https://doi.org/10.7748/nr2011.01.18.2.52.c8284

Stevens, A. (2009, August 31). How Indigenous Knowledge Is Changing Natural Hazard Mitigation. *Emergency Management*. Retrieved from https://www.govtech.com/em/disaster/Indigenous-Knowledge-Natural-Hazard-Mitigation.html

William, R., Techniques, C. O., Methods, E., Instruction, L. C., Materials, M., Evaluation, P., … Projects, S. (1999). *Observing Classroom Processes in Project-Based Learning Using Multimedia: A Tool for Evaluators By: William R. Penuel & Barbara Means.*

Zinyeka, G. (2013). Onwu and Mosimege on "Indigenous Knowledge Systems and Science and Technology Education: A Dialogue" Some Remaining Issues. *Greener Journal of Educational Research, 3*(9), 432–437.

Theory and Practice of Computation – Nishizaki et al (eds)
© 2021 Taylor & Francis Group, London, ISBN 978-0-367-41473-3

A reflective extension of a FRP language and its applications

Takuo Watanabe
Department of Computer Science, Tokyo Institute of Technology, China

ABSTRACT: This paper introduces a reflective extension of a functional reactive programming language designed for resource-constrained embedded systems. Using the extension, a program module can observe or modify its execution process via time-varying values that are connected to the internal of the metalevel of the module. Thus reflective operations are also reactive and described in a declarative manner. An example shows how the mechanism can realize an adaptive runtime that reduces the power consumption of small robots.

Keywords: Functional Reactive Programming, reflection, embedded systems

1 INTRODUCTION

Functional Reactive Programming (FRP) (Elliott & Hudak 1997; Pembeci, Nilsson, & Hager 2002; Hudak, Courtney, Nilsson, & Peterson 2003; Czaplicki & Chong 2013) is a programming paradigm for reactive systems based on the functional (declarative) abstractions of time-varying values and events. FRP has been actively studied and recognized to be promising for developing various kinds of reactive systems including robots (Pembeci, Nilsson, & Hager 2002; Hudak, Courtney, Nilsson, & Peterson 2003). This suggests that FRP can be useful for other embedded systems in general. However, with a few exceptions, the majority of the FRP (especially pure-FRP[1]) languages and systems developed so far require rich runtime resources. Hence, it is virtually impossible to run such systems on resource-constrained platforms (*e.g.*, microcontrollers).

To resolve such situation, we developed a pure-FRP language named Emfrp for small-scale embedded systems (Sawada &Watanabe 2016). The term small-scale here indicates that the target platforms of this language are not powerful enough to run conventional operating systems such as Linux. The memory footprint of typical compiled code of Emfrp is small enough to be deployed on resource-constrained devices such as microcontrollers.

An Emfrp program can be represented as a directed acyclic graph (DAG) whose nodes and edges respectively correspond to time-varying values and their dependencies. The DAG is constructed at compile-time and never change at runtime. Although this static construction guarantees the predictability of the amount of the runtime memory, it loses the flexibility of realizing adaptive behaviors at runtime.

To provide a certain degree of flexibility and adaptability to the statically designed runtime system of the language, we designed a reflection mechanism for Emfrp and discuss its use in advance of actual implementation (Watanabe & Sawada 2017). The proposed mechanism can provide a high-level and controlled access to the internal of the language runtime via time-varying values. The distinctive characteristic of our approach is that the reflective operations are also reactive.

There have been several works focusing on developing resource-constrained systems using (functional) reactive approaches (Kaiabachev, Taha, & Zhu 2007; Sant'Anna,

---

1. FRP based on purely functional languages

Ierusalimschy, & Rodriguez 2015; Helbling & Guyer 2016). However, to the best of the author's knowledge, there are no (functional) reactive systems/languages that provide reflective features.

This paper briefly describes the reflective extension and presents it usage in small robot programming. An example shows that reflection is beneficial for describing adaptable behaviors even in resource-constrained devices.

The rest of the paper is organized as follows. Section 2 briefly describes the Emfrp. Section 3 introduces the reflective extension to the language and Section 4 presents an application to small robots. Finally Section 5 concludes the paper.

2 OVERVIEW OF EMFRP

This section presents the basic (non-reflective) features of Emfrp with an example followed by the execution model of the language.

2.1 *Basics*

An Emfrp program consists of one or more *modules*. Listing 1.1 is an example module for a simple robot controller[2]. It is intended to be run on Pololu Zumo 32U4 Robot[3] (Figure 1), a palm-sized tracked robot having two motors with rotation encoders and inertial sensors (accelerometer and gyroscope). The robot is solely controlled by an on-board ATmega32U4 (8-bit AVR microcontroller with 32KB flash memory and 2.5KB RAM). The controller reads data from the gyroscope to detect when the robot is being rotated. It controls the pair of motors to cancel the rotation. As a result, the robot keeps its direction.

An Emfrp module definition consists of a single module header followed by one or more type, constant, function or node definitions used in the module. In Listing 1.1, the module header (lines 1–6) defines the module name (RotResist), then declares two input nodes (gyroZ and t) and two output nodes (motorL and motorR), and specifies the library module (Std) used in this module.

The rest of the module (lines 8–28) contains the definitions of three constants (maxSpeed, kp and ka), one function (motorSpeed) and five nodes (dt, angle, turn, motorL, and motorR). A node definition looks like

$$\text{node } n = e \text{ or node init}[c] \, n = e$$

where n is the node name and e is an expression that describes the (time-varying) value of the node. The optional init[c] specifies the constant c as the initial value of the node. Note that if e contains another node name m, we say that n refers to m and hence n depends on m. While the value of m changes over time, the value of n varies also.

Nodes are classified into three kinds: *input*, *output* and *internal*. Each input or output node has a connection to an external device (or a system entity), while an internal node has no such connection. In the example, gyroZ and t are input nodes connected to the gyroscope and system clock, respectively. Their values represent the current motion data and time. The internal nodes dt (line 18), angle (line 21) and turn (line 24) respectively express the time difference (elapsed time from the last *iteration*), the angle of the current turn, and the speed of the motor.

The definition of dt has an expression t@last, which refers to the value of t at the *previous moment* — the value evaluated in the previous *iteration* (See Section 2.2).

---

2. This example is adapted from an existing example for Pololu Zumo 32U4 Robot. https://github.com/pololu/zumo-32u4-arduino-library
3. https://www.pololu.com/category/170/zumo-32u4-robot

```
 1 module RotResist
 2 in  gyroZ : Int,        # gyroscope (z-axis)
 3      t(0)  : Int        # current time (usec)
 4 out motorL : Int,       # left motor
 5     motorR : Int        # right motor
 6 use Std
 7
 8 # This function is used to constrain the speed of the motors to be
 9 # between -maxSpeed and maxSpeed
10 const maxSpeed = 400
11 fun motorSpeed(s) = min(max(s, -maxSpeed), maxSpeed)
12
13 # PD-control parameters
14 const kp = 11930465 / 1000
15 const kd = 8
16
17 # time difference
18 node dt = t - t@last
19
20 # Calculates the angle to turn from the gyroscope data and dt
21 node angle = gyroZ * dt * 14680064 / 17578125
22
23 # Calculates the turning speed using a simple PD-control method.
24 node turn = motorSpeed(-angle / kp - gyroZ / kd)
25
26 # Controls the motors
27 node motorL = -turn
28 node motorR = turn
```

Listing 1.1. Rotation Resistant Robot Controller

Figure 1. Zumo 32U4.

2.2 Execution model

An Emfrp module can be represented as a directed graph whose nodes and edges correspond to time-varying values and their dependencies respectively. Figure 2 shows the graph representation of Listing 1.1, which consists of seven nodes and eight edges.

We categorize the edges (dependencies) into two kinds: *past* and *present*. A past edge from node *m* to *n* means that *n* has *m* @last in its definition. A present edge from node *m* to *n*, in contrast, means that *n* directly refers to *m*. In Figure 2, the dotted arrow line from t to dt is the past edge. All other edges are present.

By removing the past edges from the graph representation of an arbitrary Emfrp program, we should obtain a directed-acyclic graph (DAG). The topological sorting on the DAG gives a sequence of the nodes. For Figure 2, we have: gyroZ, t, dt, angle, turn, motorL, motorR.

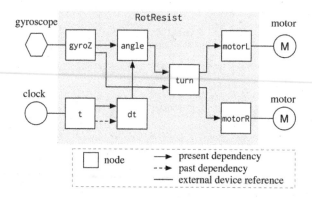

Figure 2.   Graph Representation of Listing 1.1.

The Emfrp runtime system updates the values of the nodes by repeatedly evaluating the elements of the sequence. We call a single evaluation cycle an *iteration*. The order of updates (scheduling) in an iteration must obey the partial order determined by the above mentioned DAG.

The value of n@last is the value of *n* in the last iteration. At the first iteration, where no nodes have their previous values, n@last refers to the initial value *c* specified with init[c] in the definition of *n*. In this example, since t is an input node, its initial value is specified at the header section of the module (line 3).

The Emfrp compiler translates a module definition into a platform-independent C program that repeatedly updates the values of nodes. The generated code is usually linked with some platform-dependent code (runtime system) to be deployed on an actual device.

3   REFLECTIVE EXTENSION

To provide a high-level representation of the Emfrp runtime system, we introduce the notion of *metamodule* that governs an application level (base-level) module. Figure 3 depicts the concept. A metamodule contains at least one input node (inWorld) and one output node (out-World), each of which represents an intermediate state of its corresponding base-level module.

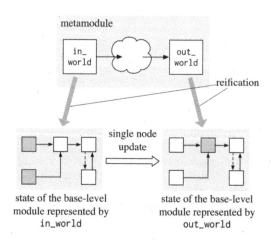

Figure 3.   Metamodule.

```
1  module VanillaMeta
2  in  inWorld  : World
3  out outWorld : World
4  use Reflect
5
6  node outWorld = {
7    (xs, ys) = inWorld
8    if isEmpty[Node](xs)
9    # Finishes a single base-level iteration
10   then (ys, empty[Node]())
11   # Updates a base-level node
12   else {
13     (x, xs') = dequeue[Node](xs)
14     (n, p, c, e) = x
15     eval(e, xs', ys) of:
16       # Updates the current value of the node
17       Just(v) -> (xs', enqueue[Node](ys, (n, c, v, e)))
18       # Does not update if evaluation fails
19       Nothing -> (xs', enqueue[Node](ys, (n, p, c, e)))
20   }
21 }
```

Listing 1.2. Vanilla Metamodule

Listing 1.2 shows the vanilla metamodule that expresses the basic execution model of Emfrp. Specifically, this module plays the role of the runtime function that repeatedly updates the node values.

Two nodes inWorld and outWorld represent an intermediate state of an iteration in the base-level module. The type of them (World) is defined as a pair type.

$$\text{type World} = (\text{Seq[Node], Seq[Node]})$$

where Seq[Node] is the sequence type whose element type is Node. Seq is a built-in type which requires an explicit type parameter (Node).

The elements of World respectively represent the nodes to be updated and the nodes already updated. The order of the nodes in the sequences should obey the order of the nodes in the dependency graph explained in Section 2.2. A single base-level iteration starts with (xs, empty-[Node]()) and ends with (empty[Node](), ys) where xs and ys respectively correspond to the sets of nodes before and after the iteration.

The type of reified nodes is defined as

$$\text{type Node} = (\text{String, Value, Value, Expr})$$

where String, Value and Expr are types of strings, reified data values (see next paragraph) and expressions. Thus, a node is represented as a quadruple (n, p, c, e) where n, p, c and e are the name, the last (previous) value, the current value, and the expression (RHS of the definition) of the node respectively. Values of the type Value represent base-level values of any data types.

Upon a successful update of a node, the previous *current* value of the node becomes the new *last* value and the evaluated value becomes the new *current* value (line 17 in Listing 1.2). If the evaluation of the node fails, the current state of the node is just used as the result (line 19 in Listing 1.2)

4   EXAMPLE: CONTROLLING THE POWER CONSUMPTION OF A ROBOT

This section describes an example using the reflective extension. The example, also runs on Zumo 32U4 Robot, uses the accelerometer to detect whether the robot is on a slanted surface. If it is on such a surface, then it turns itself to face uphill. It also uses the motor-rotation encoders to avoid rolling down the surface. Listing 1.3 show the controller module of the robot.

```
 1 module FaceUphill
 2 in  accX     : Int, # accelerometer (x-axis)
 3     accY     : Int, # accelerometer (y-axis)
 4     encL     : Int, # left motor rotation encoder
 5     encR     : Int  # right motor rotation encoder
 6 out motorL   : Int, # left motor
 7     motorR   : Int, # right motor
 8     needsTurn : Bool = meta(isBusy)
 9               # Connected to isBusy of the metamodule
10 use Std
11 meta AdaptiveSpeedMeta
12
13 # This function is used to constrain the speed of the motors to be
14 # between -maxSpeed and maxSpeed
15 const maxSpeed = 150
16 fun motorSpeed(s) = min(max(s, -maxSpeed), maxSpeed)
17
18 # True iff the robot is on a slanted surface. (incline of more than 5 degrees)
19 node init[False] needsTurn = accX * accX + accY * accY > 1427 * 1427
20
21 # Calculates the turning speed from the y-axis value of the accelerometer.
22 # It will be 0 if the incline is not significant.
23 node turn = if needsTurn then accY / 16 else 0
24
25 # Calculates the forwarding speed from the encoder values.
26 node forward = -(encL + encR)
27
28 node motorL = motorSpeed(forward - turn)
29 node motorR = motorSpeed(forward + turn)
```

Listing 1.3. Facing Uphill Robot

In line 8 of this example, a Boolean output node needsTurn is declared to be related to meta(isBusy). The notation expresses that the value of needsTurn can also be referred as the value of isBusy of the metamodule AdaptiveSpeedMeta (Listing 1.4). This *inter-level node connection* is the central mechanism of the reflective extension. Figure 4 depicts the structure of the entire program. The upper and lower lower parts of the figure show the metamodule and the base module respectively.

The metamodule AdaptiveSpeedMeta has an extra input node isBusy and an extra output node iterSleepMs as well as inWorld and outWorld. As described above, isBusy refers to the

```
 1 module AdaptiveSpeedMeta
 2 in  inWorld      : World,
 3     isBusy       : Bool   # busyness of the base-level
 4 out outWorld     : World,
 5     iterSleepMs : Int    # sleep time between iterations
 6 use Reflect
 7
 8 # Counts the iterations. Resets to 0 when detecting the falling edge of isBusy.
 9 node init[0] count =
10   if !isBusy && isBusy@last  # falling edge
11   then 0
12   else count@last + 1
13
14 # Keeps full-speed iterations while isBusy or 1000 iterations after isBusy becomes False.
15 # After that, 10ms sleep is inserted at each iteration.
16 node iterSleepMs = if isBusy || count < 1000 then 0 else 10
17
18 # Same as VanillaMeta
19 node outWorld = ...
```

Listing 1.4. Metamodule controlling the computation speed

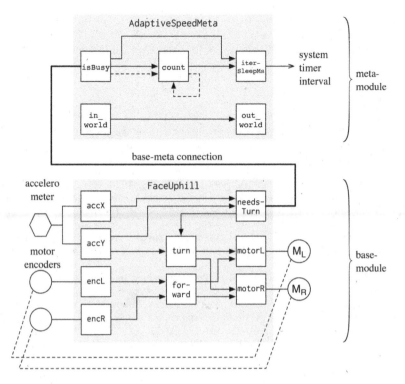

Figure 4.   Structure of FaceUphill.

value of the node needsTurn in the base-level. The node iterSleepMs represents the sleep time between iterations. The larger the value of the node is, the slower the execution of the system becomes and the smaller the power consumption will be.

In this example, while the robot is on a level plane, it will slow itself down and lower the power consumption by sleeping 10ms between iterations. Once the robot finds itself on a slanted plane, needsTurn in the base-level module becomes true. This implies that isBusy is true in the metamodule because of the inter-level node connection. Thus the system runs as fast as possible by changing the value of iterSleepMs to zero and keeps the state for 1000 iterations.

The runtime system used with this metamodule should insert specified sleep time between iterations. Such behavior can be implemented for example by inserting a sentence [language=C]usleep(iterSleepMs * 1000); to the place before the invocation of the iteration process.

5   CONCLUDING REMARK

This paper presents a simple reflective extension for Emfrp and its usage in small robot programming. The main purpose of introducing reflection is to provide a certain degree of flexibility and adaptability to the statically designed runtime system of the language.

The proposed reflection mechanism opens up the internals of a runtime system via nodes (time-varying values) connected to nodes in the metamodules. The mechanism based on the *inter-level node connection* can be classified as a behavioral reflection in a sense that the base-level module can modify its own behavior by affecting the execution of the metamodule via connected nodes.

The future research direction should focus on the investigation of the use of the proposed reflection mechanism as well as performance evaluation.

ACKNOWLEDGMENT

The author thanks the anonymous reviewers for their detailed comments. This work is supported in part by JSPS KAKENHI Grant No. 18K11236.

REFERENCES

Czaplicki, E. & S. Chong (2013). Asynchronous functional reactive programming for GUIs. In *34th ACM SIGPLAN Conference on Programming Language Design and Implementation (PLDI 2013)*, pp. 411–422. ACM.

Elliott, C. & P. Hudak (1997). Functional reactive animation. In *2nd ACM SIGPLAN International Conference on Functional Programming (ICFP 1997)*, pp. 263–273. ACM.

Helbling, C. & S. Z. Guyer (2016, Sep.). Juniper: A functional reactive programming language for the Arduino. In *4th International Workshop on Functional Art, Music, Modelling, and Design (FARM 2016)*, pp. 8–16. ACM.

Hudak, P., A. Courtney, H. Nilsson, & J. Peterson (2003). Arrows, robots, and functional reactive programming. In *Advanced Functional Programming*, Volume 2638 of *Lecture Notes in Computer Science*, pp. 159–187. Springer-Verlag.

Kaiabachev, R., W. Taha, & A. Zhu (2007). E-FRP with priorities. In *7th ACM/IEEE International Conference on Embedded Software (EMSOFT 2007)*, pp. 221–230.

Pembeci, I., H. Nilsson, & G. Hager (2002). Functional reactive robotics: An exercise in principled integration of domain-specific languages. In *4th International Conrefernce on Principles and Practice of Declarative Programming (PPDP 2002)*, pp. 168–179. ACM.

Sant'Anna, F., R. Ierusalimschy, & N. Rodriguez (2015). Structured synchronous reactive programming with Céu. In *14th International Conference on Modularity (Modularity 2015)*, pp. 29–40. ACM.

Sawada, K. & T. Watanabe (2016, Mar.). Emfrp: A functional reactive programming language for small-scale embedded systems. In *MODULARITY Companion 2016: Companion Proceedings of the 15th International Conference on Modularity*, pp. 36–44. ACM.

Watanabe, T. & K. Sawada (2017, Apr.). Towards reflection in an FRP language for small-scale embedded systems. In *2nd Workshop on Live Adaptation of Software SYstems (LASSY 2017), Companion to the 1st International Conference on the Art, Science and Engineering of Programming (Programming 2017)*, pp. 10: 1–10:6. ACM.

Theory and Practice of Computation – Nishizaki et al (eds)
© 2021 Taylor & Francis Group, London, ISBN 978-0-367-41473-3

Workflow models for integrated disease surveillance and response systems

J.C.L. Lopez & M.J. Bayuga
Service Science and Software Engineering Lab, Department of Computer Science, College of Engineering, University of the Philippines, Philippines

R.A. Juayong
Department of Computer Science, College of Engineering, University of the Philippines, Philippines

J. Malinao
Headstart Business Solutions, Inc

J. Caro
Department of Computer Science, College of Engineering, University of the Philippines, Philippines

M. Tee
Department of Physiology and Department of Medicine, College of Medicine, University of the Philippines Manila, Philippines

ABSTRACT: The Philippine Integrated Disease Surveillance and Response(PIDSR) is a disease surveillance system meant to improve the government's ability to combat epidemics. Currently, the publicly-available documentation of PIDSR's workflows in disease surveillance and response remain unstructured that induces difficulty in analyzing its overall profile and efficiency. In this research, we use a multidimensional model called the Robustness Diagram with Loop and Time Controls(RDLT) to capture this profile and describe the efficiency of such surveillance and response systems. The model can represent all workflow dimensions, i.e. resource, process, and case. As an initial step, we create other workflow models representing parts of the system under one or two of these dimensions. Thereafter, we map these preliminary models to RDLT components to build an entire profile of such disease surveillance and response systems. We shall focus on vector-borne diseases to illustrate our proposed models and strategies.

1 INTRODUCTION

As a developing country, the Philippines is home to disease vectors such as the Anopheles mosquito that is responsible for Malaria and the Aedes mosquito that is responsible for Dengue Fever. With widespread poverty and a population in the country, the threat of an epidemic is one of the biggest issues of public health [Worldometer, 2019]. The Philippine Integrated Disease Surveillance and Response (PIDSR) system was established to improve disease surveillance systems in the Philippines. The system integrates existing surveillance and response activities at all levels into one system to strengthen the capacity of local government units to detect and respond to epidemics. But how efficient is this disease surveillance and response system? Although there is a publicly-available PIDSR Manual of Procedures, there is a grave lack of formal models that profile and analyze its efficiency. The concepts, tasks, roles, cases, and procedures therein remain understudied because their relevant information remained unstructured within the Manual, and in general, in the pool of available literature.

Because the system attempts to involve multiple levels of the government to deal, manage, and monitor diseases, it is quite natural that there is a challenge in getting a holistic view of the system. In this research, we create a multidimensional workflow model that is suitable and adaptable enough to overcome this challenge. Thereafter, we analyze this multidimensional model to get a better understanding of the system in its entirety, and possibly yield previously unknown insights to help increase its efficiency. In this research, we use the multidimensional model called Robustness Diagrams with Loop and Time Controls(RDLT) as proposed by (Malinao et al., [Malinao et al., 2016]) to model systems such as PIDSR.

1.1 Philippine integrated disease surveillance and response

The PIDSR system is a streamlined combination of multiple disease surveillance systems that is meant to act as one unified disease surveillance system for the country. A project of the Department of Health (DOH), the country's disease surveillance and response systems from the barangay to the national level are fully integrated into one [National Epidemiology Center, 2014]. The data generated by the PIDSR system is used in the government's public health decision-making. It is also used in the detection of and response to epidemics. The system's various, multi-level processes such as the flow of health information and instructions on proper case reporting, are all detailed within the PIDSR Manual of Procedures. It is this surveillance system that the researchers wish to study by making a multidimensional workflow model of the system's processes, specifically how it handles the reporting of Vector-borne Diseases(VBD).

Definition 1.1 *A vector-borne disease [Philippines, 2018] is a human illness caused by vectors - living organisms such as mosquitoes, flies, and ticks that transfer infectious diseases.*

Definition 1.2 *PIDSR categories for diseases, syndromes, and events are the following:*

- **Category I** - *Category I diseases, syndromes, and events are those that are considered urgent and require immediate action. They are listed as Immediately Notifiable in the PIDSR system, and each Category I case has its own PIDSR Case Investigation Form.*
- **Category II** - *Category II diseases, syndromes, and events are those that are considered less urgent. They are listed as Weekly Notifiable in the PIDSR system, and share the same PIDSR Case Report Forms. [Department of Health, 2014]*

Figure 1 shows the workflow graphs from the PIDSR Manual of Procedures that our RDLT model will be compared to.

1.2 Robustness diagrams with loop and time controls

A RDLT is an extension of the Robustness Diagram of the UML. RDLTs allow users to create a model using all three workflow dimensions, i.e. process, resource, and case.

Definition 1.3 *A RDLT is a multidimensional workflow that is expressed as a graph with the following components and attributes:*

- **Vertex Set.** *Each vertex of a RDLT is either a (1) boundary or entity object, or a (2) controller. Vertices which are objects represent real-world components of the system. In particular, boundary objects represent components that interact with the system's environment. Meanwhile, entity objects are internal components of the system that do not interact with the environment but with other components within the system. Lastly, controllers represent tasks that are performed by the components of the system. Every object has an attribute M that has a value of either 0 or 1.*
- **Arc Set.** *Each arc can connect two vertices in the RDLT as long as these two vertices are not both objects. That is, every arc (x, y) in the RDLT has endpoints x and y where x can be a boundary/entity object while y is a controller(or vice versa), or x and y are both controllers. Furthermore, each arc has the following attributes:*

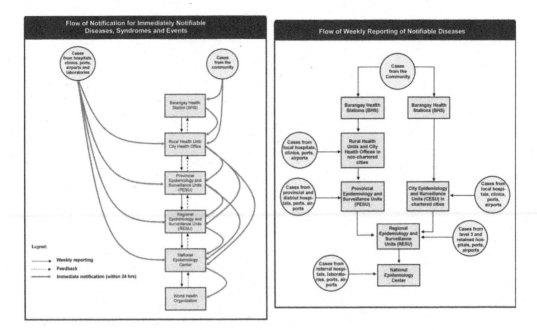

Figure 1. The PIDSR system's flow of notification for immediately notifiable (Category I) and weekly notifiable (Category II) diseases [National Epidemiology Center, 2014].

- *Constraint Label. Each arc has a label $c \in \Sigma \cup \{\varepsilon\}$ where Σ is a finite non-empty set of symbols representing constraints that must be satisfied for a traversal to be made possible on the arcs, and ε is the empty string(i.e. no constraint).*
- *Maximum Number of Traversals allowed on the arc. This marks the maximum allowable times that algorithms can perform traversals on the arc.*

Remark 1.1 *A traversal can only happen on an arc (x, y) if all the required constraints in the arcs connecting from the parents of y are satisfied and their maximum number of traversals had not been exceeded.*

- *Time Vector. This vector stores the time step/s that an algorithm traverses the arc or the first time the algorithm considers the (another) possible traversal on the arc.*
- *Reset-bound subsystem(RBS). A RBS is a subgraph of the RDLT induced by an object x in the subgraph whose attribute $M(x) = 1$ and every vertex y where (x, y) is an arc of the RDLT where its constraint label is ε. Traversing an arc (s, t) in the RDLT where s is a vertex in the RBS and t is not in the RBS shall reset to 0 all values of the time vectors in all the arcs within the RBS.*

Two arcs (a, b) and (x, y) of the RDLT are type-alike if any of the following hold:

- *a and x are outside of RBS Q and $b = y$ where b is inside Q,*
- *$a = x$ where a is inside of RBS Q and both b and y are outside Q,*
- *$b = x$ where b is inside of RBS Q and both a and y are outside Q,*
- *a, b, x, y are all inside or all outside of RBS Q.*

Shown in Figure 2 is a sample RDLT showing the components and attributes mentioned in Definition 1.3. It has one entity object x1, one boundary object x2, and controllers y1, y2, y3, y4, and y5. Note that there is an RBS in the RDLT that is induced by x2 where $M(x2) = 1$, along with the controllers y4 and y5. Here, the arcs (y1, x2) and (y2, x2) are type-alike. However, (y1,x2) and (x2, y4) are not type-alike. If there was an arc (y5, x2), then (y1, x2) and (y5, x2) are not type-alike.

143

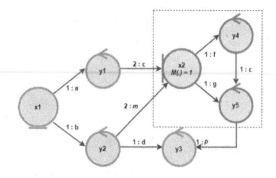

Figure 2.   An example of a RDLT.

Remark 1.2 *Type-alike arcs with target node x shall solely be the ones considered for the traversal of x. That is, their constraint labels shall solely be the required set of constraints included in the consideration of traversal of x.*

We can analyze the profile and efficiency of real-world systems by the RDLT workflows by identifying the latter's interesting features, i.e. *Points-of-Interest*(POIs) and neigborhood structures thereof. A POI in the RDLT can be the following: (a) **Point of Delay (POD)**, (b) **Point of Re-entry**, and (c) **Point of Synchronization(POS)**. In particular, these POIs are identified using the following concepts and neighborhood,

- **POD** x - A vertex x in the RDLT is a **POD** if there are type-alike arcs (a, x) and (b, x) where their constraint labels are different.
- *Elementary path* - A path from one vertex to another without repetition of any other vertices involve in the said path.
- *Antecedent Set of vertex x* - The set of vertices along an elementary path from a source vertex to the vertex x in the RDLT.
- *Consequent Set of vertex x* - The set of vertices along a path from x to some vertex y where y is a member of the antecedent set of x, and excluding the vertices in the antecedent set.
- *Looping Arc of POD x* - An arc (a, b) is a *looping arc* used by POD x in the RDLT if b is in the antecedent set of x and either (1) a is in the consequent set of x, or (2) the consequent set of x is null.
- **POR** v of POD x - A vertex y is a **POR** for a POD x if (u, v) is a looping arc used by x.
- **POS** u of POD x - A vertex u is a **POS** for a POD x if either (1) u is a source along a path towards x, or (2) u is a POS of x.

2   RELATED WORK

2.1   *Manual of procedures for the Philippine integrated disease surveillance and response [National epidemiology Center, 2014][National Epidemiology center, 2014]*

The PIDSR Manual of Procedures provides a descriptive discussion of the processes for the notification and reporting of cases through multiple government levels. There are three(3) PIDSR forms that are used in the recording and reporting of these cases which are the following: (1) Weekly Notifiable Disease Report (WNDR) Summary Page, (2) Case Investigation Forms for Category I diseases/syndromes, and (3) Case Report Forms for Category II diseases/syndromes.

The set of VBDs are mostly placed under the Category II diseases, although some specific examples like Malaria can be treated as Category I. Nonetheless, we shall study both categories though their PIDSR forms in this research. These forms collect data on the human health at each level by Disease Surveillance Coordinators(DSC) and Disease Surveillance

Officers(DSO). They are submitted to higher level Disease Reporting Units(DRUs) which are responsible for consolidating, analyzing, and interpreting the data. Thereafter, they will create their own reports that will be disseminated to other units and passed on to the higher level of the government.

However, the Manual is observed to be inconsistent in its terms and vague in its descriptions. For example, in Manual of Procedures for the Philippine Integrated Disease Surveillance and Response, 3rd Edition, Volume 2, Annex 4(dated 2014), Malaria is clearly listed under the list of Category II (weekly notifiable) diseases, syndromes, and health events. Note that Category II diseases only use Case Report Forms. However, Malaria has its own PIDSR Case Investigation Form as of 2016, which are only meant for Category I diseases, syndromes, and events, and yet the 3rd edition of the manual is the latest version publicly available despite already being outdated. Another example would be the role of the DSC. Specifically, Section 3.3.2 of the Manual states that DSCs are detailed as staff assigned to hospitals, private clinics, and rural health units(and thus are part of the municipal level and above), and yet in Section 4.3.1 are present in the Barangay Health Station as part of disease surveillance at the Barangay level. The Manual is thus unsuitable for analysis on its own without other sources of information to clarify these inconsistent information of its processes.

2.2 Building RDF Models of Multidisciplinary Data Sets under the One Health Framework in the Philippine Setting [Dela Rosa et al., 2018]

This work introduces the use of the Resource Description Framework(RDF) data model to represent multidisciplinary data sets from the Philippines under the One Health(OH) framework. In particular, the data sets came from the CHED-PCARI project on One Health itself, and PIDSR's Manual of Procedures and Forms inclusive of the Case Investigation Form, Case Report Form, and Weekly Notifiable Disease Report Summary Page. Thereafter, these heterogeneous data sets were profiled, converted into RDF graphs, integrated into a unified model, and analyzed. Specifically, this research had the following contributions for the One Health Framework:

1. conversion of the unstructured data sets to structured ones, i.e. RDF triplestores and models,
2. defining a Predicate Construction Model on top of the RDF components to be able to capture relationships of the various information in the OH Framework in terms of (a) VBDs and possible environmental and human contributors and symptoms thereof, and (b) the governmental structures that collect and manage these information from the barangay to the national level.
3. a method of unifying heterogenous datasets into one RDF Model using the Predicate Construction Model and a graph algorithm to find transitive associations across these datasets.
4. performing model analysis to gain new insights of the One Health Framework using the unified RDF Model.

The RDF models that were developed in the aforementioned research put more focus on the conceptual aspects of the field structures and relationships, it still lacks information on the efficiency of the One Health Framework. In terms of workflow dimensions as well as the definition of RDF models, there is emphasis on building the the models with the resource dimension. Meanwhile, there is not so much emphasis on describing the models, or implying some derived workflows from these models, to check the efficiency of the One Health Framework within the domain of the data sets under study.

3 METHODOLOGY

In order to build the RDLT model of the PIDSR, we shall create preliminary models whose definition and construction focus on one or two dimensions. In particular, we shall build the Component, Activity, and Sequence Diagrams of the UML. The Component Diagram shall show the PIDSR resources and the relationships that are or can be established using Manual. These resources can be medical and health professionals, government officials, or any artifacts of the system where data is recorded such as the PIDSR Forms, database tables where data is stored, etc. Meanwhile, the Activity Diagram highlights the use of the process and case dimensions of workflows. More specifically, these Diagrams can capture sequential or iterated tasks, and forks where conditions impose or produce branching processes ending up to some specific case. Lastly, Sequence Diagrams highlights the resource and process dimensions, and are by practice, showing only one case per diagram. In these diagrams, the resources are shown as actors where their roles, managed inputs and outputs, and interactions are drawn on their respective lifelines. Creating all these models provides the researchers a means to first express parts of the PIDSR in simpler forms before its corresponding multidimensional workflow is built. This research shall show how the components of these preliminary models are mapped to our chosen multidimensial workflow, i.e. the RDLT. For illustration purposes of this methodology, we shall apply it to profile and analyze the PIDSR specifications on the responsibilities of the DSCs and DSOs, as well as the workflow on the notification of reporting of disease cases in the Manual's Section 3 and 4, respectively.

Shown below are the transformation process of the elements of the Component, Activity, and Sequence Diagrams to elements of the RDLT.

Rationale of the Component Diagram to RDLT Transformation

Table 1 summarizes the transformation of component diagram elements to RDLT elements. For Transformations 1 and 2, *Components* could be represented either by Boundary objects or Entity objects depending entirely on the connections of its ports. If a *Component* has a *port* connecting it to a port on a different level, then it may be represented as a Boundary object. If all of its interactions stay within the same level, the Component becomes an Entity object.

For Transformation 3, the *Port* element does not contain enough information in order to be fully translated to an RDLT graph. They show us the existence of a controller connecting two boundary or entity objects, but the actual content and number of controllers must be determined by the Activity and Sequence diagrams.

For Transformation 4, *Components with dependencies* are not converted into the RDLT as their dependencies mean they become a part of the controllers owned by the components they are dependent on.

Table 1 . Transformation of component diagram to RDLT.

Component Diagram Elements[OMG,2011]	RDLT Elements
1. Components (Without dependencies, with ports that connect it to other components in a separate level	Boundary Object
2. Components (Without dependencies, without ports that connect it to other components in a separate level	Entity Object
3. Ports	A controller that connects a boundary/entity object to another. The description of the controller is determined by the description of the message between its equivalent objects in the sequence diagrams. The number of controllers in between them is determined by the number of messages between objects.
4. Components (with dependencies)	Not converted. Components with dependencies instead become a part of a controller's description, which is described by the Activity and Sequence Diagrams
5. Packages	Preserved upon conversion to RDLT

Finally, for Transformation 5, upon the creation of the Component Diagrams, the components were all grouped together by *packages* according to the government level each component was a part of as dictated by the PIDSR Manual of Procedures. These groupings were preserved upon conversion to RDLT, which is used to distinguish Boundary from Entity objects.

Rationale of the Activity Diagram to RDLT Transformation

Table 2 summarizes the transformation of activity diagram elements to RDLT elements. For Transformations 1, 2, and 12, in Activity Diagrams, the*Initial, Activity final*, and*Flow final* nodes indicate the starting and ending points of the diagram, as well as a point where the process ends respectively. This is represented within the RDLT as a source node for the initial node, and sink nodes to represent both the activity final and flow final nodes.

For Transformation 3, an *activity* represents the activities that occur in the process. It is represented in the RDLT by two vertices. A controller that states the activity being carried out, and the boundary or entity object that owns the process. Note that all activities in the initial models follow the format of**Noun A** verb **Noun B**, with **Noun A** naming the boundary or entity object that owns the diagram and the verb naming the controller that **Noun A** owns. **Noun B** names the object acted upon and directed to, which is either a component with a dependency named in the component diagrams and thus is not converted to the RDLT, or a component that becomes a boundary or entity object that the outgoing arc in the RDLT conversion points towards.

The rationale behind Transformations 4, 7, and 10 is self-evident, and the activity diagrams used in this research did not make use of *Swimlanes*. They have merely been added to the table for the use of future research.

For Transformations 5, 6, 8 and 9, *Forks* split one flow into multiple, denoting the beginning of parallel activity while Joins mark the end of parallel activity, needing all flows going into it to reach it before processing continues. On the other hand, *Decisions* also have one flow entering and multiple flows leaving, but use conditions to determine which of the multiple leaving flows is used by the process. Conversely, *Merges* do not need every flow entering it in order to continue processing, only one or more incoming based on any conditions. These

Table 2. Activity to RDLT.

Activity Diagram Elements[Ambler, 2005]	RDLT Elements
1. Initial Node	Source
2. Activity Final Node	Entity Object
3. Activity	Two vertices, a controller that states the process and the boundary/entity object that owns the process
4. Flow/Edge	Arcs
5. Forks	Two or more arcs with only ε constraint moving out of the same vertex
6. Join	Two or more arcs with only ε constraints moving into the same vertex
7. Condition	Constraints
8. Decision	A controller and multiple outgoing arcs (one arc for each possible decision) leaving it, but with constraints on all leaving arcs.s
9. Merge	Multiple arcs entering the same vertex, but with Σ constraints on at least one arc.
10. Partitions/Swimlanes	Ownership arcs
11. Sub-activity indicator	Controllers labeled as abstraction labels
12. Flow final node	Sink
13. Use Case	Maximal Substructure
Special Rule: Two activities separated by a decision with the same Noun B on the same level	A boundary or entity object with two arcs whose constraints are the conditions of the decision

are represented in the RDLTs through the use of constraints, with multiple arcs entering and leaving the same vertex with only ε constraints recreating the parallel processing of *Forks* and *Joins*, while a controller with constraints on its multiple outgoing arcs and a vertex with multiple incoming arcs with Σ constraints on at least one arc simulate the decision making processing of *Decisions* and *Merges*.

For Transformations 11 and 13, Sub-Activity Indicators are activities that are described by a seperate, more finely detailed activity diagram, while Use Cases are notes indicating the use case being invoked by the Activity Diagram. These are represented in the RDLT by use of a controller that acts as an abstraction label that is described by a more finely detailed RDLT, while Use Cases are described by Maximal Substructures within the RDLT.

Finally, the special rule was introduced because there were certain activities with exactly the same descriptions that nevertheless were on separate flows from a previous decision node. Due to the conversion to RDLTs making it so that the arcs enter the same boundary/entity object, the special rule ensures that the exclusiveness of the flows is preserved.

Rationale of the Sequence Diagram to RDLT Transformation

Table 3 summarizes the transformation of sequence diagram elements to RDLT elements. For Transformations 1, 2, and 5, *Objects* being represented by Boundary or Entity objects is self-evident, as well as representing *Activation Occurrences* by Ownership arcs. The Chronological order of the messages in a sequence diagram is also preserved upon conversion to RDLT, due to the RDLT's timesteps.

For Transformation 3, *Messages* represent communication between objects, and thus are represented in the RDLT as controllers. Their source and target objects are also preserved.

Finally, Transformation 4 is similar to how *Use Cases* are represented in Activity Diagram conversions, represented by Maximal Substructures within the RDLT.

3.1 Elements of an RDLT not captured by the initial models

Within RDLTs, the Reset Bound Subsystem (RBS) is an element of the model that has no direct equivalent within Component, Activity, and Sequence Diagrams. While similar structures are certainly possible within Activity and Sequence Diagrams, it is not possible to simulate the resetting of the timestep of the RBS back to 0 upon traversal into or out of the subgraph.

3.2 Conversion to a higher level RDLT

Due to the size of a multi-level, finely-detailed RDLT graph, this research also includes a set of rules for converting the graph to a higher level form that is more conducive for validation by comparison to the PIDSR Manual of Procedures' graphs, as well as for ease of reading. When we are converting the RDLT graph to a higher level form, the following vertices of the lower level model are preserved:

- Boundary and Entity objects who own or have incoming controllers that involve passing PIDSR forms
- Controllers with multiple arcs owned by DSCs or DSO boundary objects whose constraints involve city type (chartered/component)

Table 3.   Sequence Diagram to RDLT.

Sequence Diagram Elements[and]	RDLT Elements
1. Objects	Boundary or Entity objects. Boundaries are the levels determined by Component Diagrams
2. Methods (or Activation occurrence)	Ownership Arcs
3. Messages	Controllers. Source and target objects are preserved
4. Use cases	Maximal substructure
5. Chronological orders	Preserved within conversion to RDLT via timestep or T

- Controllers with multiple arcs whose constraints involve Disease Category (I/II)
- Controllers that involve filling out PIDSR forms
- Controllers that involve passing PIDSR forms to the next government level
- Controllers whose description involves the patient approaching any health staff (hospital staff, RHU staff, BHWs. etc)
- Controllers involving disease cases reported by email or telephone

While these rules may appear similar to vertex simplification [Malinao, 2016], they preserve the forks that may occur within the government levels such as in disease case classification or that of city type classification.

3.3 Calculating the time of delay

Upon creating the RDLT model and analyzing it for any Points of Interests, we calculate the *time of delay* [Malinao, 2016] of every POD present in the model. The **time of delay** of a POD x is the maximum of the differences of the path length from its ancestor POS a to the a' farthest descendant u where (u, x) is in the RDLT and the constraint label of (u, x) is in Σ, against the path length from a to its nearest descendant v where (v, x) is in the RDLT, and where (u, x) and (v, x) are type-alike arcs. The value of this maximum (plus one) tells the waiting time, i.e. delay, that the POD is reachable from a after reaching v the first time.

4 RESULTS AND DISCUSSION

4.1 The diagram with loop and time robustness controls

Each of the initial component, activity, and sequence diagrams were grouped by the levels they described, then converted into a single RDLT graph per level using the transformation rules in order to create the multi-level model. The following are keywords used in our modelling in this study.

- *Accommodated* is defined as either having enough space or available resources(e.g. medicine) to support the patient.
- Although all Philippine cities are classified as Chartered Cities, Batas Pambansa 51 classifies cities into Highly Urbanized Cities(HUC) and Component Cities, and BP 337 refines it further. We are using HUC for the PIDSR's definition of Chartered Cities(income of at least 30M pesos and 150k inhabitants), everything else is a Component City.
- A DRU is declared *silent* when it has not submitted PIDSR reports, including a failure to report on no cases, for 2 or more weeks
- A technical assistance visit is when an expert from a higher level of the surveillance system visits a health facility to provide on-site mentoring, training, and addresses issues of implementation of PIDSR

Note that the sub-activity indicator in the initial Activity Diagram models does not feature a 1:1 conversion within the RDLT. This was done on purpose, as the researchers identified another similar repeating structure attached to what would have been the Abstraction Label, and included it in the Abstraction Label in order to simplify the resulting RDLT graph.

4.2 Points of interest in the RDLT

Fpr the RDLT models(see Appendix) that we constructed for all the levels of the government, i.e. barangay, municipal, provincial, regional, and national, we conducted an analysis by identifying POIs. This analysis revealed two PODs present in the RDLT, vertices p22 and a6. The following are the pertinent information for the two PODS.

- **Antecedent Set for POD** p22: [x1, x2, x3, x4, x5, x6, x7, x8, x9, x14, x15, x16, x17, x18, x19, x20, x21, x22, x23, x24, x25, m1, m2, m3, m4, m5, m6, m7, m8, m9, m10, m11, m12, m13, m14, m15, m16, m17, m18, m19, m20, m21, m22, m23, m24,m25, m26, p1, p2, p3, p4, p5, p6, p7, p8, p9, p10, p18, p19, p20, p21, p22, p23]
- **Consequent Set for POD** p22: NULL;
- **Looping arc for POD** p22: NULL
- **Antecedent Set for POD** a6: [x1, x2, x3, x4, x5, x6, x7, x8, x9, x14, x15, x16, x17, x18, x19, x20, x21, x22, x23, x24, x25, m1, m2, m3, m4, m5, m6, m7, m8, m9, m10, m11, m12, m13, m14, m15, m16, m17, m18, m19, m20, m21, m22, m23, m24, m25, m26, p1, p2, p3, p4, p5, p6, p7, p8, p9, p10, p11, p12, p13, p14, p15, p16, p17, p18, p19, p20, p21, p22, p23]
- **Consequent Set for POD** a6: [a7, a8, a9, a10, a11, a12, a13, a14, a15}
- **Looping arcs for POD** a6: [(a11, a4), (a15, a4)]
- **POR** for p22: NULL; **POS** for p22: x1
- **POR** for a6: a4, **POS** for a6: [x1, a4]
- For p22: **Longest path from POS** to p21 = 21, **Shortest path from POS** to p21 = 12; **Time of Delay** of p22 = 10 time steps
- For a6: (a) with a4 as POS, all Time of Delays of a6 are 1 since there's only a single way to move from a4 to a5, and (b) with x1 as POS, Longest path from POS to a5 = 31, Shortest path from POS to a5 = 13, Time of Delay of a6 = 19 time steps

Within the RDLT, the point of delays, vertices p22 and a6, are the nodes "maintain forms" within the Provincial level and "conducts assessment to verify and resolve silent status" within the Abstraction Label A(see Appendix), which is contained in the Provincial and Regional levels respectively. These points make sense as points of delays, since a DSO at the Provincial level having to stop to encode handwritten or physical forms into a computer first delays maintaining a file of PIDSR forms at p22, while the "conducts assessment" node is able to be accessed by two completely different DSOs, the DSO of the Provincial Level and the DSO of the Regional level, hampering the efficiency of the workflow by a significant amount having a total time of delay of 19 time steps.

4.3 The high-level RDLT

The following graph is the high level form of the multi-level RDLT, formed when the conversion rules stated in the methodology were applied to the multi-level RDLT.

RDLT Comparison to the Manual's Workflow for Immediately Notifiable Cases

As can be seen in Figure 3, there is a clear path through the RDLT that simulates the Immediately Notifiable Workflow graph from the Manual as reflected in Figure 1.1. There are however, some minor but significant differences between the graphs. The RDLT does not feature a direct line from a case into the Regional level and National level, as the Flow of Notification of cases in Section 4 of the PIDSR Manual of Procedures did not contain any details about hospitals that would report to the DSO of the RESU at the Regional level nor the PHSID at the National level, instead listing how DSCs located in hospitals reported either to the municipal or the provincial levels. There is also a significant difference when it comes to who receives immediate notification from an immediately notifiable case from the Barangay level. The Manual's graph shows that upon receiving an Immediately Notifiable case, the Barangay Health Station reports only to the Rural Health Unit, whereas in the graph, the Barangay Health Station reports simultaneously to the PESU, the RESU, and the NEC at the Provincial, Regional, and National levels respectively just like the Rural Health Unit once it receives an Immediately Notifiable case. This disparity is caused by Section 4.2 of the manual dictating that any DRU that receives an Immediately Notifiable Case must report simultaneously to the PESU, RESU, and NEC, and the manual specifically lists a Barangay Health Station as a DRU in its Glossary, while not giving any special

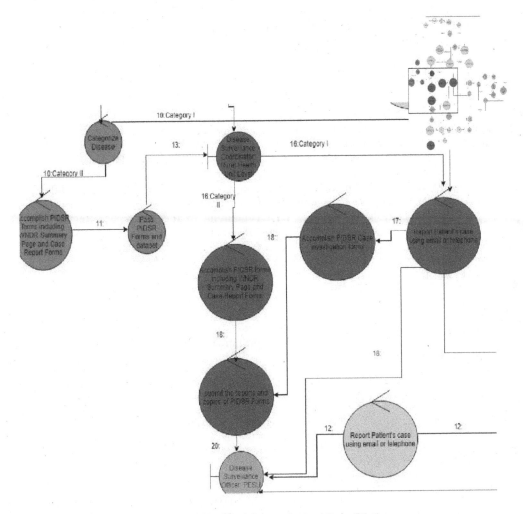

Figure 3. A part of RDLT graph's higher level form to be used for validation.

instructions otherwise for a Barangay Health Station in Section 4.2. It is possible the Manual itself has missing information on who a Barangay Health Station must report to, or there is an error in the Manual's graph itself.

RDLT Comparison to the Manual's Workflow for Weekly Notifiable Cases

As can be seen in Figure 3, there is also a clear path through the RDLT that simulates the Weekly Notifiable Workflow graph from the Manual as reflected in Figure 1, including the split in flow between Chartered and Component cities and their rejoining at the Regional level before reporting to the National level. Similar to the comparisons to the workflow for Weekly Notifiable cases however, there is no way for a case from a hospital to enter the Regional and National levels as the Flow of Notification of cases in Section 4 of the PIDSR Manual of Procedures did not contain any details about how the hospitals could report a case directly to those levels. There are also no descriptions that differentiate local, provincial, district, level 3, and referral hospitals from each other nor their roles within the PIDSR system, with Section 4 of the Manual only referring to "hospitals" as an umbrella term for them all. Thus, there is still a small though significant difference between the Weekly Notifiable workflow graph and the RDLT when it comes to the sources of Weekly Notifiable cases.

5 CONCLUSIONS

This research led to the creation of component, activity, and sequence diagrams that describe the notification workflow of disease cases in the PIDSR system, as well as create an RDLT model from these initial models that captured the same workflow. This RDLT model revealed inconsistencies between the specific instructions stated in sections 4.2 and 4.3 of the Manual and the workflows depicted at the end of section 4. One example of this is the difference between the RDLT model and the Immediately Notifiable Graph's handling of a Category I case at the barangay level, with the RDLT pointing to the Provincial, Regional, and National level as instructed in section 4.2 while the Immediately Notifiable Graph points to the Municipal level. This model also revealed two point of delays in the PIDSR workflow: The process of maintaining forms by the DSO (PESU), and the process of verifying and resolving silent DRUs by all DSOs.

REFERENCES

Ambler, S. (2005). UML 2 Activity Diagrams: An Agile Introduction. Available at http://agilemodeling. com/artifacts/activityDiagram.htm.

Dela Rosa, M., Tesoro, N., Malinao, J., Canseco, R., Codera, A., and Caro., J. (2018). Building RDF models of Multidisciplinary Data Sets under the One Health Framework in the Philippine Setting.

Malinao, J. (2016). On Building Multidimensional Workflow Models for Complex Systems Modelling (Dissertation).

Malinao, J., Judex, F., Selke, T., and Zucker, G. et al. (2016). Rdlts for system modelling and scenario extraction with energy system applications.

National Epidemiology Center (2014). Manual of procedures for the philippine integrated disease surveillance and response, 3rd ed, vol 1.

National Epidemiology center (2014). Manual of procedures for the philippine integrated disease surveillance and response, 3rd ed, vol 2.

OMG (2011). OMG Unified Modeling Language (OMG UML), Superstructure, Version 2.4.1 Object Management Group (Technical report, Object Management Group). Available at http://www.uml-sysml.org/documentation/uml-2.1.2-superstructure-5.8mo/at_download/file.

Philippines, W. R. O. (2018). Communicable Diseases. Available at http://www.wpro.who. int/philippines/areas/communicable_diseases/mvp/en/.

Rosenberg, D. and Scott, K. (2001). Sequence diagrams: One step at a time. Available at http://www8.tfe.umu.se/courses/systemteknik/Doit/UML/Sequence

Worldometer (2019). Philippines Population (2019) - Worldometer. Available at http://www.worldometers.info/world-population/philippines-population/.

APPENDIX

Figure 4 shows one sample of the RDLT models that represent the various levels of government. These were converted to the higher level model seen in Figure 3. The full set of models is available in the following website: https://sites.google.com/up.edu.ph/upd-dcs-s3lab/publications

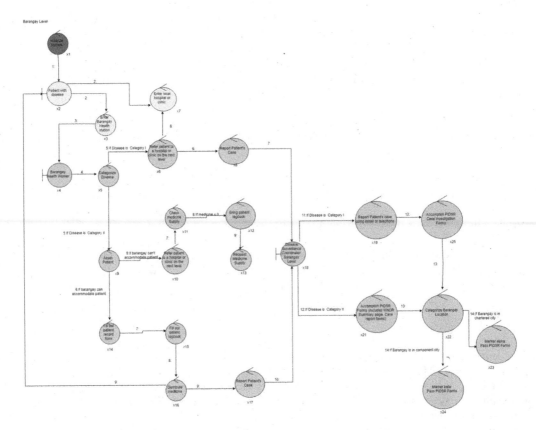

Figure 4.   The RDLT graph's Barangay level.

Theory and Practice of Computation – Nishizaki et al (eds)
© 2021 Taylor & Francis Group, London, ISBN 978-0-367-41473-3

Preservation of well-handledness in Petri net reduction rules

Louis Anthony Agong & H.N. Adorna
University of the Philippines, Diliman, Philippines

ABSTRACT: As systems evolve and increase in complexity, it can be difficult to maintain and verify some properties modelled by Petri nets. Reduction rules and its equivalent synthesis rules can be used for the transformation of the evolving Petri nets. It is preferable that these transformation methods preserve properties as they evolve. This paper prove that well-handledness can be strongly preserved after applying some reduction rules to an arbitrary Petri net. Additionaly, this result provide a way to partially verify well-handledness in Petri nets.

1 INTRODUCTION

A Petri net (PN) (Petri, 1962) is a mathematical tool that can be used in modelling concurrent systems. Petri net's properties are wildly studied and different subclasses of are explored. Behavioral properties such as liveness, boundedness and safeness are used to describe the dynamic behavior of the Petri net. The analysis of the properties depend primarily on the initial marking defined on the Petri net. Structural properties like free choiceness and well-handledness are used describe the static structure of the Petri net. The structural properties are not marking dependent and focus more on the topological structure of Petri nets. This paper focuses more on the well-handledness of arbitrary Petri nets.

Well-handledness can be used to describe a 'good' Petri net. In a well-handled Petri net two parallel paths initiated by an OR-split must always joined by an OR-join. Additionally, two parallel paths initiated by an AND-split must always joined by an AND-join. A well-handled Petri net implies a number of good properties. For example a strongly connected Petri net is a well-formed Petri net (van der Aalst, 1996).

Given a large and complex PN, it is useful for us to have methods that can reduce the complex PN to a more simpler and smaller one. These reduction methods have a gives us advantages. For instance, verification of liveness and boundedness are decidable but also EXPSPACE-hard (Cheng et al., 1995). By reducing a Petri net using reduction rules that can preserve liveness and boundedness properties we can save a lot of computational resources for verification of both resources. Murata introduced six (6) reduction rules which preserve liveness, safeness and boundedness (Murata, 1989).

In this paper we will show that of the set of reduction rules presented by Murata (Murata, 1989) preserves well-handledness on the general class of Petri nets.

The paper is organized as follows. In section 2 we present the definitions and preliminary concepts that will be used in our proof. Section 3 details our proof for strong preservation of well-handledness. Section 4 concludes the paper and presets possible future work.

2 PRELIMINARIES

We define in the section the definitions that will be used in our proof.

Definition 1.1 (Petri Net) *A Petri Net is a 3-tuple, (P, T, F) where:*
*– $P = \{p_1, p_2, \ldots, p_m\}$ is a finite set of **places***

- $T = \{t_1, t_2, \ldots, t_m\}$ is a finite set of **transitions**
- $F \subseteq (P \times T) \cup (T \times P)$ is a set of arcs
- In particular, a place p is called an **input (output)** place of a transition t if there exists an arc from p to t (t to p) and a transition t is called an **input (output)** transition of a place p if there exists an arc from t to p (p to t)
- We define a **marking** as a mapping $P \rightarrow N$ that represents the number of tokens on the places. If p is a place, p^k represents k tokens on p. Thus, $M = p_1{}^{k_1} p_2{}^{k_2} \cdots p_n{}^{k_n}$ represents a marking that has k_i tokens in place p_i for $i = 1, 2, \ldots, n$.
- ${}^\bullet t = \{p | (p, t) \in F\}$ = the set of input places of t
- $t^\bullet = \{p | (t, p) \in F\}$ = the set of output places of t
- ${}^\bullet p = \{t | (t, p) \in F\}$ = the set of input transitions of p
- $p^\bullet = \{t | (p, t) \in F\}$ = the set of output transitions of p
- If $G \subseteq P \cup T$ then $G^\bullet = \cup_{g \in G} g^\bullet$
- If $G \subseteq P \cup T$ then ${}^\bullet G = \cup_{g \in G} {}^\bullet g$

Definition 1.2 (Transition Firing Rules) *A transition t is said to be **enabled** if each input place p of t contains at least one token. An enabled transition may or may not fire. If a transition t **fires**, then t consumes one token from each input place p of t and produces one token for each output place of t. Let M_a, M_b be two markings. We denote $M_a \xrightarrow{t} M_b$ the fact that transition t is enabled by marking M_a and firing t results in the marking M_b. Let $M_1, M_2, \ldots M_n$ be markings and $\sigma = t_1 t_2 \ldots t_{n-1}$ as sequence of transitions. We denote $t' \in \sigma$ if $t' \in \{t_1, \ldots, t_{n-1}\}$. We denote $M_1 \xrightarrow{\sigma} M_n$ the fact that $M_1 \xrightarrow{t_1} M_2 \xrightarrow{t_2} \cdots \xrightarrow{t_{n-1}} M_n$.*

Definition 1.3 (Path) *Let $N = (P, T, F)$ be a Petri net. A **path** C from a node n_1 leading to n_k is the sequence $(n_1, n_2, \ldots, n_k)$ such that $(n_i, n_{i+1}) \in F$ for $1 \leq i \leq k$. A path C is an **elementary path** if for every two nodes n_i and n_j, $i \neq j \Rightarrow n_i \neq n_j$. We define an operator α such that $\alpha(C) = \{n_1, n_2, \ldots, n_k\}$.*

Definition 1.4 (Petri net transformation rule) *A **Petri net transformation rule** or simply **transformation rule**, usually denoted by ϕ, is a binary relation on the class of Petri nets. It is fully described by the conditions of application under which it can be applied to the source Petri net, and the construction algorithm that is applied to the source Petri net to form the target Petri net. A transformation rule is called a **reduction rule** (**synthesis rule**) if the number of nodes of the target net is strictly smaller (larger) than the source net.*

Definition 1.5 (Applicable transformation rule) *Let N, N' be two Petri nets. A transformation rule ϕ is said to be **applicable** to N if applying ϕ to N results in N'. We denote this by $(N, N') \in \phi$*

Definition 1.6 (Workflow property preservation) *Let N be a Petri net and let ψ be a Petri net property. $N \models \psi$ denotes the fact that N satisfies property ψ. If, for all Petri nets N, N' where $(N, N') \in \phi$ and $N \models \psi \Rightarrow N' \models \psi$, then ϕ is said to **preserve** ψ. If the reverse is also true, i.e. $N' \models \psi \Rightarrow N \models \psi$, then ϕ is said to **strongly preserve** ψ.*

Definition 1.7 (Well-handledness) *Let N be a Petri net. Let x and y be a pair of nodes in N such that one node is a place and the other one is a transition. The node pair is said to satisfy the **well-handledness property** if for any pair of elementary paths C_1 and C_2 leading from x to y,*

$$\alpha(C_1) \cap \alpha(C_2) = \{x, y\} \Rightarrow C_1 = C_2.$$

*A Petri net N is **well-handled** if all pair of nodes x and y satisfy the well-handledness property.*

3   PRESERVATION OF THE WELL-HANDLEDNESS REDUCTION RULES

In this section we show that the well-handledness of Petri nets are strongly preserved after applying the reduction rules presented in Murata's paper (Murata, 1989). For each reduction rule we show that N is well-handled if and only if the constructed $\overline{N}$ is well-handled.

3.1 Fusion of series places (ϕ_{FSP})

Conditions on N:

$$\exists t \in T \tag{1}$$

$$\exists p \in P\backslash\{i, o\} \tag{2}$$

$$^{\bullet}t = \{p\} \tag{3}$$

$$p^{\bullet} = \{t\} \tag{4}$$

$$t^{\bullet} = \{p_{out}\} \tag{5}$$

Construction of $\overline{N}$

$$\overline{P} := P\backslash\{p\} \tag{6}$$

$$\overline{T} := T\backslash\{t\} \tag{7}$$

$$\overline{F} := (F\backslash(\{ (p, t), (t, p_{out})\} \cup (^{\bullet}p \times \{p\}))) \cup (^{\bullet}p \times \{p_{out}\}) \tag{8}$$

Figure 1 Provides the graphical depiction of ϕ_{FSP}.

Proposition 1 ϕ_{FSP} *strongly preserves well-handledness.*

Proof. ($\Rightarrow$) Let N be a well-handled Petri net that satisfies the conditions for ϕ_{FSP}. Note that all node pairs in N satisfy the well-handledness property. Let x and y be nodes in $\overline{N}$ of different types. Suppose C_1 and C_2 are elementary paths in $\overline{N}$ from x leading to y. Furthermore, suppose $\alpha(C_1) \cap \alpha(C_2) = \{x, y\}$.

Case 1: Suppose p_{out} is not on either path. Then C_1 and C_2 are both in N which implies that $C_1 = C_2$ since N is well-handled.

Case 2: Suppose p_{out} is on either path.

Case 2.1: Suppose $p_{out} = x$. Then C_1 and C_2 are both in N which implies that $C_1 = C_2$ since N is well-handled.

Case 2.2: Suppose $p_{out} = y$. Suppose $C_1 \neq C_2$. Let $C_1 = xa_1a_2\ldots a_ip_{out}$ and $C_2 = xb_1b_2\ldots b_jp_{out}$. Note that the existence of C_1 and C_2 in $\overline{N}$ implies the existence of the two paths $C_3 = xa_1a_2\ldots a_ip$ and $C_4 = xb_1b_2\ldots b_jp$ in N. We also have $C_3 \neq C_4$ and $\alpha(C_3) \cap \alpha(C_4) = \{x, p\}$ which is a contradiction since N is well-handled. Therefore $C_1 = C_2$.

Case 2.3: Suppose $p_{out} \neq x$ and $p_{out} \neq y$. Suppose $C_1 \neq C_2$. Without loss of generality, assume that p_{out} in path C_1. Let $C_1 = xa_1\ldots a_ip_{out}b_1\ldots b_jy$. Since C_1 exists in $\overline{N}$ there is a path C_3 in N such that $C_3 = xa_1\ldots a_iptp_{out}b_1\ldots b_jy$. Hence we have $C_2 \neq C_3$ where $\alpha(C_2) \cap \alpha(C_3) = \{x, y\}$ which is a contradiction. Therefore $C_1 = C_2$.

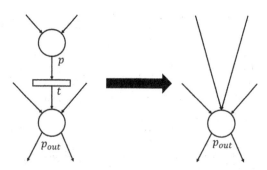

Figure 1. ϕ_{FSP} example from (Murata, 1989).

($\Leftarrow$) Let $\overline{N}$ be a well-handled Petri net that is constructed after applying ϕ_{FSP}. Let x and y be nodes in N of different types. Suppose C_1 and C_2 are elementary paths in N from x leading to y. Furthermore, suppose $\alpha(C_1) \cap \alpha(C_2) = \{x,y\}$.

Case 1: Suppose $y = p$. Moreover, suppose $C_1 \neq C_2$. Then we after applying ϕ_{FSP} we could obtain two paths C_3 and C_4 such that $\alpha(C_3) \cap \alpha(C_4) = \{x, p_{out}\}$ and $C_3 \neq C_4$ which is a contradiction. Therefore $C_1 = C_2$.

Case 2: Suppose t is on either path.

Case 2.1 Suppose $x = t$. Since $\alpha(C_1) \cap \alpha(C_2) = \{x,y\}$ we have $y = p_{out}$ which implies that $C_1 = C2$.

Case 2.2 Suppose $y = t$. Since $\alpha(C_1) \cap \alpha(C_2) = \{x,y\}$ we have $x = p$ which implies that $C_1 = C2$.

Case 2.3 Suppose $x \neq t$ and $y \neq t$. Suppose $C_1 \neq C_2$. Without loss of generality, assume that t is in path C_1. Let $C_1 = a_1 \ldots a_i p t p_{out} b_1 \ldots b_j$. After applying the rule ϕ_{FSP} we could find a path C_3 in $\overline{N}$ such that $C_3 = a_1 \ldots a_i p_{out} b_1 \ldots b_j$. Hence we have $C_2 \neq C_3$ where $\alpha(C_2) \cap \alpha(C_3) = \{x,y\}$ which is a contradiction. Therefore $C_1 = C_2$.

Case 3: Otherwise, C_1 and C_2 are both in $\overline{N}$ which implies that $C_1 = C_2$ since $\overline{N}$ is well-handled.

Therefore N is well-handled.

3.2 Fusion of series transitions (ϕ_{FST})

Conditions on N:

$$\exists p \in P \backslash \{i, o\} \tag{9}$$

$$\exists t \in T \tag{10}$$

$$p^\bullet = \{t\} \tag{11}$$

$$^\bullet t = \{p\} \tag{12}$$

$$^\bullet p = \{t_{in}\} \tag{13}$$

Construction of $\overline{N}$

$$\overline{P} := P \backslash \{p\} \tag{14}$$

$$\overline{T} := T \backslash \{t\} \tag{15}$$

$$\overline{F} := (F \backslash (\{ (p,t), (t_{in}, p)\} \cup (\{t\} \times t^\bullet))) \cup (\{t_{in}\} \times t^\bullet) \tag{16}$$

Figure 2 Depicts graphically ϕ_{FST}..

Proposition 2 ϕ_{FST} strongly preserves well-handledness.

Proof. ($\Rightarrow$) Let N be a well-handled Petri net that satisfies the conditions for ϕ_{FST}. Note that all node pairs in N satisfy the well-handledness property. Let x and y be nodes in $\overline{N}$ of different types. Suppose C_1 and C_2 are elementary paths in $\overline{N}$ from x leading to y. Furthermore, suppose $\alpha(C_1) \cap \alpha(C_2) = \{x,y\}$.

Case 1: Suppose t_{in} is not on either path. Then C_1 and C_2 are both in N which implies that $C_1 = C_2$ since N is well-handled.

Case 2: Suppose t_{in} is on either path.

Case 2.1: Suppose $t_{in} = x$. Suppose $C_1 \neq C_2$. Let $C_1 = t_{in} a_1 a_2 \ldots a_i y$ and $C_2 = t_{in} b_1 b_2 \ldots b_j y$. Note that the existence of C_1 and C_2 in $\overline{N}$ implies the existence of the two paths $C_3 = t a_1 a_2 \ldots a_i y$ and $C_4 = t b_1 b_2 \ldots b_j y$ in N. We also have $C_3 \neq C_4$ and $\alpha(C_3) \cap \alpha(C_4) = \{p, y\}$ which is a contradiction since N is well-handled. Therefore $C_1 = C_2$.

Case 2.2: Suppose $t_{in} = y$. Then C_1 and C_2 are both in N which implies that $C_1 = C_2$ since N is well-handled.

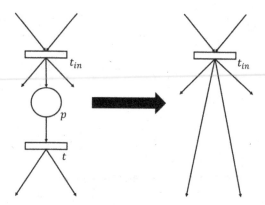

Figure 2. ϕ_{FST} example from (Murata, 1989).

Case 2.3: Suppose $t_{in} \neq x$ and $t_{in} \neq y$. Suppose $C_1 \neq C_2$. Without loss of generality, assume that t_{in} in path C_1. Let $C_1 = xa_1 \ldots a_i t_{in} b_1 \ldots b_j y$. Since C_1 exists in $\overline{N}$ there is a path C_3 in N such that $C_3 = xa_1 \ldots a_i t_{in} p t b_1 \ldots b_j y$. Hence we have $C_2 \neq C_3$ where $\alpha(C_2) \cap \alpha(C_3) = \{x, y\}$ which is a contradiction. Therefore $C_1 = C_2$.

($\Leftarrow$) Let $\overline{N}$ be a well-handled Petri net that is constructed after applying ϕ_{FST}. Let x and y be nodes in N of different types. Suppose C_1 and C_2 are elementary paths in N from x leading to y. Furthermore, suppose $\alpha(C_1) \cap \alpha(C_2) = \{x, y\}$.

Case 1: Suppose $x = t$. Moreover, suppose $C_1 \neq C_2$. Then we after applying ϕ_{FST} we could obtain two paths C_3 and C_4 such that $\alpha(C_3) \cap \alpha(C_4) = \{t_{in}, y\}$ and $C_3 \neq C_4$ which is a contradiction. Therefore $C_1 = C_2$.

Case 2: Suppose p is on either path.

Case 2.1 Suppose $x = p$. Since $\alpha(C_1) \cap \alpha(C_2) = \{x, y\}$ we have $y = t$ which implies that $C_1 = C2$.

Case 2.2 Suppose $y = p$. Since $\alpha(C_1) \cap \alpha(C_2) = \{x, y\}$ we have $x = t_{in}$ which implies that $C_1 = C2$.

Case 2.3 Suppose $x \neq p$ and $y \neq p$. Suppose $C_1 \neq C_2$. Without loss of generality, assume that p is in path C_1. Let $C_1 = a_1 \ldots a_i t_{in} p t b_1 \ldots b_j$. After applying the rule ϕ_{FST} we could find a path C_3 in $\overline{N}$ such that $C_3 = a_1 \ldots a_i t_{in} b_1 \ldots b_j$. Hence we have $C_2 \neq C_3$ where $\alpha(C_2) \cap \alpha(C_3) = \{x, y\}$ which is a contradiction. Therefore $C_1 = C_2$.

Case 3: Otherwise, C_1 and C_2 are both in $\overline{N}$ which implies that $C_1 = C_2$ since $\overline{N}$ is well-handled.

Therefore N is well-handled.

3.3   Fusion of parallel places (ϕ_{FPP})

Conditions on N:

$$\exists p, p' \in P \setminus \{i, o\} \tag{17}$$

$$^{\bullet}p = {}^{\bullet}p' = \{t_{in}\} \tag{18}$$

$$p^{\bullet} = p'^{\bullet} = \{t_{out}\} \tag{19}$$

Construction of $\overline{N}$

$$\overline{P} := P \setminus \{p'\} \tag{20}$$

$$\overline{T} := T \tag{21}$$

$$\overline{F} := F \setminus \{(t_{in}, p'), (p', t_{out})\} \tag{22}$$

158

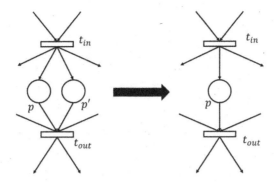

Figure 3. ϕ_{FPP} example from (Murata, 1989).

Figure 3 Is the graphical representation of ϕ_{FPP}.

Proposition 3 ϕ_{FPP} *strongly preserves well-handledness.*

Proof. ($\Rightarrow$) Let N be a well-handled Petri net that satisfies the conditions for ϕ_{FPP}. Note that all node pairs in N satisfy the well-handledness property. Moreover, applying ϕ_{FPP} to produce $\overline{N}$ remove the paths containing p' and no new paths are created. Thus, all node pairs in $\overline{N}$ still satisfy the well-handledness property. Therefore $\overline{N}$ is well-handled.

($\Leftarrow$) Suppose the constructed Petri net $\overline{N}$ is well-handled. Let x and y be nodes in N of different types. Suppose C_1 and C_2 are elementary paths in N from x leading to y. Furthermore, suppose $\alpha(C_1) \cap \alpha(C_2) = \{x, y\}$. We need to show that $C_1 = C_2$

Case 1: Suppose p' is not on either path. Then the two paths are in $\overline{N}$ which implies that $C_1 = C_2$ since $\overline{N}$ is well-handled.

Case 2: Suppose p' is on either path.

Case 2.1: Suppose $p' = x$. Note that p' has only one outgoing transition t_{out}. Hence, the arc (x, t_{out}) are both on C_1 and C_2. But $\alpha(C_1) \cap \alpha(C_2) = \{x, y\}$. Hence, $t_{out} = y$ and therefore $C_1 = C_2$.

Case 2.2: Suppose $p' = y$. Note that p' has only one ingoing transition t_{in}. Hence, the arc (t_{in}, y) are both on C_1 and C_2. But $\alpha(C_1) \cap \alpha(C_2) = \{x, y\}$. Hence, $t_{in} = x$ and therefore $C_1 = C_2$.

Case 2.3: Suppose $p' \neq x$ and $p' \neq y$. Without loss of generality, suppose p' is in C_1. We could also find a path C_3 that is similar to C_1 except that we replace p' by p defined in one of the conditions for ϕ_{FPP}. Note that C_2 and C_3 are both paths in $\overline{N}$. Moreover, $\alpha(C_2) \cap \alpha(C_3) = \{x, y\}$ which implies that $C_2 = C_3$ which is a contradiction since C_3 has an additional node p. Therefore this case cannot happen.

Therefore, N is well-handled.

3.4 *Fusion of parallel transitions* (ϕ_{FPT})

Conditions on N:

$$\exists t, t' \in T \tag{23}$$

$$^\bullet t = {}^\bullet t' = \{p_{in}\} \tag{24}$$

$$t^\bullet = t'^\bullet = \{p_{out}\} \tag{25}$$

Construction of $\overline{N}$

$$\overline{P} := P \tag{26}$$

$$\overline{T} := T \backslash \{t'\} \tag{27}$$

$$\overline{F} := F \backslash \{(p_{in}, t'), (t', p_{out})\} \tag{28}$$

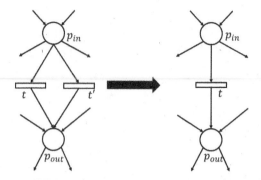

Figure 4. ϕ_{FPT} example from (Murata, 1989).

The graphical depiction of ϕ_{FPT} is shown in Figure 4.

Proposition 4 ϕ_{FPT} *strongly preserves well-handledness.*

Proof. ($\Rightarrow$) Let N be a well-handled Petri net that satisfies the conditions for ϕ_{FPT}. Note that all node pairs in N satisfy the well-handledness property. Moreover, applying ϕ_{FPP} to produce $\overline{N}$ remove the paths containing t' and no new paths are created. Thus, all node pairs in $\overline{N}$ still satisfy the well-handledness property. Therefore $\overline{N}$ is well-handled.

($\Leftarrow$) Suppose the constructed Petri net $\overline{N}$ is well-handled. Let x and y be nodes in N of different types. Suppose C_1 and C_2 are elementary paths in N from x leading to y. Furthermore, suppose $a(C_1) \cap a(C_2) = \{x, y\}$. We need to show that $C_1 = C_2$

Case 1: Suppose t' is not on either path. Then the two paths are in $\overline{N}$ which implies that $C_1 = C_2$ since $\overline{N}$ is well-handled.

Case 2: Suppose t' is on either path.

Case 2.1: Suppose $t' = x$. Note that t' has only one outgoing place p_{out}. Hence, the arc (x, p_{out}) are both on C_1 and C_2. But $a(C_1) \cap a(C_2) = \{x, y\}$. Hence, $p_{out} = y$ and therefore $C_1 = C_2$.

Case 2.2: Suppose $t' = y$. Note that t' has only one ingoing place p_{in}. Hence, the arc (p_{in}, y) are both on C_1 and C_2. But $a(C_1) \cap a(C_2) = \{x, y\}$. Hence, $p_{in} = x$ and therefore $C_1 = C_2$.

Case 2.3: Suppose $t' \neq x$ and $t' \neq y$. Without loss of generality, suppose t' is in C_1. We could also find a path C_3 that is similar to C_1 except that we replace t' by t defined in one of the conditions for ϕ_{FPT}. Note that C_2 and C_3 are both paths in $\overline{N}$. Moreover, $a(C_2) \cap a(C_3) = \{x, y\}$ which implies that $C_2 = C_3$ which is a contradiction since C_3 has an additional node t. Therefore this case cannot happen.

Therefore, N is well-handled.

3.5 *Elimination of self-loop places (ϕ_{ESP})*

Conditions on N:

$$\exists p \in P \tag{29}$$

$$^\bullet p = p^\bullet = \{t\} \tag{30}$$

Construction of $\overline{N}$

$$\overline{P} := P \backslash \{p\} \tag{31}$$

$$\overline{T} := T \tag{32}$$

$$\overline{F} := F \backslash \{(p, t), (t, p)\} \tag{33}$$

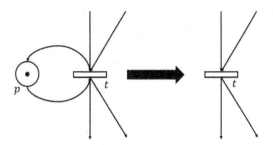

Figure 5. ϕ_{ESP} example from (Murata, 1989).

Note that to preserve liveness after reduction it is required to have at least one token present in place p. However, since our concern in this paper is the preservation of well-handledness we also consider Petri nets with places without tokens in them. (Figure 5 shows an instance of ϕ_{ESP}.)

Proposition 5 ϕ_{ESP} *strongly preserves well-handledness.*

Proof. ($\Rightarrow$) Let N be a well-handled Petri net that satisfies the conditions for ϕ_{ESP}. Note that all node pairs in N satisfy the well-handledness property. Moreover, applying ϕ_{ESP} to produce $\overline{N}$ remove the paths containing t and no new paths are created. Thus, all node pairs in $\overline{N}$ still satisfy the well-handledness property. Therefore $\overline{N}$ is well-handled.

($\Leftarrow$) Suppose the constructed net $\overline{N}$ is well-handled. Let x and y be nodes in N of different types. Suppose C_1 and C_2 are elementary paths in N from x leading to y. Furthermore, suppose $\alpha(C_1) \cap \alpha(C_2) = \{x, y\}$.

If p is not on either path, then the two paths are in $\overline{N}$ which means that $C_1 = C_2$. Note that we can only form elementary paths with p if it is on the beginning or at the end of the path. Suppose, without loss of generality, $p \in C_1$ (otherwise $p \in C_2$). We have two cases:

Case 1: Suppose $p = x$. Note that p has only one outgoing transition t. Hence, the arc (x, t) are both on C_1 and C_2. But $\alpha(C_1) \cap \alpha(C_2) = \{x, y\}$. Hence, $t = y$ and therefore $C_1 = C_2$.

Case 2: Suppose $p = y$. Note that p has only one ingoing transition t. Hence, the arc (t, y) are both on C_1 and C_2. But $\alpha(C_1) \cap \alpha(C_2) = \{x, y\}$. Hence, $t = x$ and therefore $C_1 = C_2$.

Hence, N is well-handled.

3.6 Elimination of self-loop transitions (ϕ_{EST})

Conditions on N:

$$\exists t \in T \tag{34}$$

$$^\bullet t = t^\bullet = \{p\} \tag{35}$$

Construction of $\overline{N}$

$$\overline{P} := P \tag{36}$$

$$\overline{T} := T \backslash \{t\} \tag{37}$$

$$\overline{F} := F \backslash \{(p, t), (t, p)\} \tag{38}$$

ϕ_{EST} is illustrated in Figure 6.

Proposition 6 ϕ_{EST} *strongly preserves well-handledness.*

Proof. ($\Rightarrow$) Let N be a well-handled Petri net that satisfies the conditions for ϕ_{EST}. Note that all node pairs in N satisfy the well-handledness property. Moreover, applying ϕ_{EST} to produce $\overline{N}$ remove the paths containing t and no new paths are created. Thus, all node pairs in $\overline{N}$ still satisfy the well-handledness property. Therefore $\overline{N}$ is well-handled.

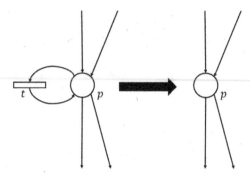

Figure 6. ϕ_{EST} example from (Murata, 1989).

($\Leftarrow$) Suppose the constructed net $\overline{N}$ is well-handled. Let x and y be nodes in N of different types. Suppose C_1 and C_2 are elementary paths in N from x leading to y. Furthermore, suppose $\alpha(C_1) \cap \alpha(C_2) = \{x, y\}$.

If t is not on either path, then the two paths are in $\overline{N}$ which means that $C_1 = C_2$. Note that we can only form elementary paths with t if it is on the beginning or at the end of the path. Suppose, without loss of generality, $t \in C_1$ (otherwise $t \in C_2$). We have two cases:

Case 1: Suppose $t = x$. Note that t has only one outgoing place p. Hence, the arc (x, p) are both on C_1 and C_2. But $\alpha(C_1) \cap \alpha(C_2) = \{x, y\}$. Hence, $p = y$ and therefore $C_1 = C_2$.

Case 2: Suppose $t = y$. Note that t has only one ingoing place p. Hence, the arc (p, y) are both on C_1 and C_2. But $\alpha(C_1) \cap \alpha(C_2) = \{x, y\}$. Hence, $p = x$ and therefore $C_1 = C_2$.

Hence, N is well-handled.

4 CONCLUSION

This paper proved that the reduction rules presented by Murata (Murata, 1989) preserves well-handledness and is applicable to the general class of Petri nets. This method is especially useful for well-handledness verification since one can significantly reduce a Petri net size before checking the property using traditional means. Synthesis rules can also be obtained from the Murata's reductions rules. With this, Petri net refinements can use the obtained synthesis rules to ensure that well-handlednes can be preserved along with liveness, boundedness, and safeness.

Desel and Esparza introduced three reduction rules that can preserve the free-choice property for arbitrary Petri nets. Moreover, they also showed that the reduction rules are complete with respect to the free-choice property (Desel and Esparza, 1995). It is desirable to also find a set of complete reduction rules with respect to well-handledness of Petri nets. Other reduction rules (Desel and Esparza, 1995, Bride et al., 2017) may be explored to see if they also preserve well-handledness in Petri nets.

REFERENCES

Bride, H., Kouchnarenko, O., and Peureux, F. (2017). Reduction of workflow nets for generalised soundess verification.

Cheng, A., Esparza, J., and Palsberg, J. (1995). Complexity results for 1-safe nets. *Theoretical Computer Science*, 147(1):117–136.

Desel, J. and Esparza, J. (1995). *Free Choice Petri Nets*. Cambridge University Press, Cambridge.

Murata, T. (1989). Petri nets: Properties, analysis and applications. *Proceedings of the IEEE*, 77(4):541–580.

Petri, C. A. (1962). *Kommunikation mit Automaten*. PhD thesis, Universitt Hamburg.

van der Aalst, V. (1996). Structural characterization of sound workflow nets. *Computing Science Report Eindhoven University of Technology*.

Theory and Practice of Computation – Nishizaki et al (eds)
© *2021 Taylor & Francis Group, London, ISBN 978-0-367-41473-3*

Gradual typing for environment calculus

Kota Inamori, Takuya Matsunaga & Shin-ya Nishizaki
Tokyo Institute of Technology, Tokyo, Japan

ABSTRACT: We can categorize programming languages into two groups; one is a group of statically-typed programming languages and another that of dynamically- typed programming languages. Gradual typing enables us to integrate static and dynamic typing in a single programming language, which was proposed by Siek et al. They formalize gradual typing as the gradually typed lambda calculus, GTLC and gives cast insertion and the internal cast calculus, which are formulated cast-insertion technique in compiling programs in a gradually-typed programming language.

In this paper, we study gradual typing in the environment calculus, a lambda calculus with first-class environment. We propose a gradually typed environment calculus, a cast internal environment calculus, and cast insertion.

1 INTRODUCTION

We can categorize programming languages into two groups; one is a group of statically-typed programming languages and another that of dynamically-typed programming languages. In the languages of the former group, typing of a program is checked in compile-time, for example, C, C++, Java, Haskell, OCaml, etc. Advantages of such languages are the following points.

- You can find some of program bugs in compile-time.
- A compiler can utilize typing information of a program source code for generating more efficient codes.
- Explicit typing information helps you to understand a program.

In the languages of the latter group, such as Lisp, Python, and Javascript, typing is checked in execution-time. Dynamically-typed languages have the following merits.

- They enable us to do prototype-programming using a dynamically-typed language more easily than the statically-typed programming languages since a programmer is not bothered by typing.
- Changing specifications, you can change a program of a dynamically-typed language more quickly than the dynamically-typed language.

Gradual typing, proposed by Siek et al. (Siek & Taha 2006)(Siek, Vitousek, Cimini, & Boyland 2015) enables us to integrate static and dynamic typing in a single programming language, and it aims to enjoy both advantages.

They propose the Gradually Typed Lambda Calculus (GTLC), the Internal Cast Calculus (ICC), and the cast insertion. The cast is a mechanism that converts a type of an expression explicitly in a typed programming language. For example, in C-like languages such as C, C++, Java, etc., the value of 1/3 is 0 since it is an integer quotation; that of 1/(double)3 0.333 · · · since it is a floating-point number's quotation. Cast enables us to make explicit type conversion. The GTLC is a language for gradual typing, in which type conversion is provided implicitly. The ICC is a language in which type conversion must be made explicitly using cast. The cast insertion is a translation of the GTLC into the ICC, which inserts casts appropriately in a program code. Their calculi are based on the simple type theory; Igarashi et al. (Igarashi, Sekiyama, & Igarashi 2017) extend their work to the gradual typing based on the polymorphic type system.

A *first-class object* is a semantic entity that can be passed to a function and returned from a function as a computation result. For example, an integer is a first-class entity in many programming languages. However, in C programming language, a pointer to a function is first-class, but a function itself is not. In functional programming languages such as Scheme and Haskell, a function can be handled as a first-class object.

An *environment* is a data structure in the programming language semantics, which is represented as a mapping of variable names to bound values. In the lambda calculus (Abadi, Cardelli, Curien, & Lévy 1991), the environment is formalized as a substitution. For example, consider a lambda-term $\lambda x.\lambda y.(x + x + y)$. If you give actual parameters 1 and 2, then we have

$$(\lambda x.\lambda y.(x + x + y))12 \rightarrow (\lambda y.(x + x + y)[x := 1])2 \rightarrow (x + x + y)[x := 1, y := 2].$$

The last term can be considered a term $(x + x + y)$ to which a substitution $[x := 1, y := 2]$ of 1 and 2 for x and y, respectively. The substition $[x := 1, y := 2]$ can be interpreted as an environment which assigns 1 and 2 to the variables x and y, respectively. In the $\lambda \sigma$-calculus (Abadi, Cardelli, Curien, & Lévy 1991), proposed by Abadi et al., an environment is formalized as a substitution not defined in the object-level. Nishizaki(Nishizaki 1995)(Nishizaki 1994)(Nishizaki & Fujii 2012) studied the first-class environment in the framework of the lambda calculus and the $\lambda \sigma$-calculus. In his calculus, he formalized the first-class environment by handling the lambda terms and the substitution uniformly.

In this paper, we study the cast insertion in the environment calculus. First, we give a Gradually Typed Environment Calculus (GTEC), in which type conversion is made implicitly. Second, we show a Cast Internal Environment Calculus (CIEC), in which type conversion is provided by cast. Third, we propose cast insertion, which is a transformation of an expression of GTEC to the one of CIEC and then discuss theoretical properties on the systems.

2 GRADUALLY TYPED ENVIRONMENT CALCULUS

The Gradually Typed Environment Calculus (GTEC) is an extension of the Gradually Typed Lambda Calculus (GTLC) proposed by Siek et al. (Siek & Taha 2006) (Siek, Vitousek, Cimini, & Boyland 2015). The GTLC is based on the simply-typed lambda calculus (Girard, Taylor, &Lafont 1989) and the dynamic type $*$ whose type is unknown statically, sometimes called *Any* type. In the GTLC and the GTEC, not only a simply-typed expression but also an expression that may or may not be typed is allowed.

Definition 1.1 (Type of GTEC) *A basic type of GTEC is either* **int** *or* **bool**:

$$B ::= \textbf{int} \,|\, \textbf{bool}.$$

A type of GTEC is defined by the following grammar:

$$T ::= B|(S \rightarrow T)|\,\textbf{env}\,(\Gamma)|\,\textbf{env}^*(\Gamma)|\,*.$$

A type $(S \rightarrow T)$ is called a function type whose domain and codomain is of type S and T, respectively. Types **env**(Γ) *and* **env**$^*(\Gamma)$ *are called a (non-dynamic) environment type and a dynamic environment type, respectively.*

A type assignment Γ of GTEC is a sequence of variable-type pairs:

$$\Gamma ::= *|\{x_1 : T_1\} \cdots \{x_n : T_n\}$$

where the variable $x_1, \ldots, x_n$ are different from each other, that is, $x_i \neq x_j$ for $i \neq j$. If $n = 0$, the type assignment is written as $\{\}$.

In the previous papers (Nishizaki 1995)(Nishizaki 1994)(Nishizaki & Fujii 2012), the order of variable-type pairs in an environment type is not cared about; $\{x : A\}\{y : B\}$ and $\{y : B\}\{x : A\}$ are identified. However, in *GTEC*, we distinguish them. If you let Γ_1 and Γ_2 be $\{x : A\}\{y : B\}$ and $\{y : B\}\{x : A\}$ respectively, Γ_1 and Γ_2 are distingueshed; a non-dynamic environment type $env(\Gamma_1)$ and a dynamic environment type $\mathbf{env}(\Gamma_2)$ are distinguished. $\mathbf{env}^*(\Gamma_1)$ and $\mathbf{env}^*(\Gamma_2)$ can be identified under type consisitensy which is proposed below: $\mathbf{env}^*(\Gamma_1) \sim \mathbf{env}^*(\Gamma_2)$.

Definition 1.2 (Term of *GTEC*) A set **Const** of constants and a set **Var** of variables are provided prior to defining a term.

Terms of *GTEC* are defined by the following grammar.

$$M ::= c \,|\, x \,|\, (\lambda x : T.M) \,|\, (MN) \,|\, id \,|\, (M/x) \cdot N \,|\, (M \circ N)$$

where $c \in$ **Const** and $x \in$ **Var**.

Term $(\lambda x : T.M)$ is a lambda abstraction where the variable x is of type T. Term (MN) is a function application. Term id is called an identity environment, which return the current environment, such as the-environment of Scheme. Term $(M/x) \cdot N$, called an environment extension, is an environment extended adding a binding of the variable x to M. Term $(M \circ N)$, called an environment composition, is an evaluation of term M under the environment value obtained from evaluation of N. The syntax of term is the same as the original environment calculi proposed by Nishizaki et al. (Nishizaki 1995) (Nishizaki 1994) (Nishizaki & Fujii 2012).

In the following, we introduce a notion of type consisitency, which is obtained by extending the one of GTLC. The type consistency $S \sim T$ means that types S and T are the same or share a shape.

Definition 1.3 (Type Consisitency of *GTEC*) Type consisitency $T_1 \sim T_2$ is a binary relation between types T_1 and T_2 defined inductively by the following rules.

$$\frac{}{* \sim T} \quad \frac{}{T \sim *} \quad \frac{}{B \sim B} \quad \frac{T_1 \sim T_3 \quad T_2 \sim T_4}{(T_1 \to T_2) \sim (T_3 \to T_4)}$$

$$\frac{}{\mathbf{env}^*(\Gamma) \sim T} \quad \frac{}{T \sim \mathbf{env}^*(\Gamma)} \quad \frac{}{\mathbf{env}(\Gamma) \sim \mathbf{env}(\Gamma)}$$

The type consistency is formalized as a kind of compatibility between types, which was proposed originally by Siek et al. (Siek, Vitousek, Cimini, & Boyland 2015). For example, $(int \to int) \sim (* \to *)$. In this paper, we represent difference of handling between $\mathbf{env}(\Gamma)$ and $\mathbf{env}^*(\Gamma)$. Let Γ_1 and Γ_2 be $\{x : A\}\{y : B\}$ and $\{y : B\}\{x : A\}$ respectively. $env(\Gamma_1) \not\sim env(\Gamma_2)$ but $\mathbf{env}^*(\Gamma_1) \sim \mathbf{env}^*(\Gamma_2)$.

Definition 1.4 (Typing Judgement and Rules) A *typing judgement* $\Gamma \vdash M : T$ *is a ternary relation among a type assignment Γ, a term M, and a type T, defined inductively by the following typing rules.*

$$\frac{\{x : T\} \in \Gamma}{\Gamma \vdash x : T} \text{ Var} \quad \frac{\{c : T\} \in \Delta}{\Gamma \vdash c : T} \text{ Const} \quad \frac{\{x : U\}\Gamma \vdash M : T}{\{x : T\}\Gamma \vdash \lambda x : S.M : S \to T} \text{ Lam}$$

$$\frac{\Gamma \vdash M : T_1 \to T_2 \quad \Gamma \vdash N : T'_1 \quad T'_1 \sim T_1}{\Gamma \vdash (MN) : T_2} \text{ App} \quad \frac{\Gamma \vdash (MN) : * \quad \Gamma \vdash M : * \quad \Gamma \vdash N : T}{\Gamma \vdash (MN) : *} \text{ AppA}$$

$$\frac{\Gamma \vdash N : \mathbf{env}(\Gamma') \quad \Gamma' \vdash M : T}{\Gamma \vdash (M \circ N) : T} \text{ Comp}$$

$$\frac{\Gamma \vdash N:\ \mathbf{env}^*(\Gamma')\ \Gamma'' = \mathrm{DynEnv\,Type}(\mathrm{Var}(M)) \sqcup \Gamma'\ \Gamma'' \vdash M:T}{\Gamma \vdash (M \circ N):T}\ \mathrm{CompA}$$

$$\frac{\Gamma \vdash M:T\ \Gamma \vdash N:\ \mathbf{env}\,(\{x:S\}\Gamma')}{\Gamma \vdash (M/x) \cdot N:\ \mathbf{env}(\{x:T\}\Gamma')}\ \mathrm{Ext} \qquad \frac{\Gamma \vdash M:T\ \Gamma \vdash N:\ \mathbf{env}^*(\{x:S\}\Gamma')}{\Gamma \vdash (M/x) \cdot N:\ \mathbf{env}^*(\{x:T\}\Gamma')}\ \mathrm{ExtA}$$

In typing judgement $\Gamma \vdash M:T$, Γ describes type information on the free variables appearring in term M and under such situation, M is of type T.

Example 1.1 We show an example of a derivation of cast insertion in $GTEC$.

$\vdash (\lambda e{:}\mathbf{env}^*(\{\}).(x \circ e))\big((1/x) \cdot id\big) \rightsquigarrow$
$\quad (\lambda e{:}\mathbf{env}^*(\{\}).x \circ (e:\mathbf{env}^*(\{\}) \Rightarrow \mathbf{env}^*(\{x{:}*\})))\big((1/x) \cdot id:\mathbf{env}(\{x{:}\mathbf{int}\}) \Rightarrow *\big)$
$\quad :*$
(CApp)

$\vdash \lambda e{:}\mathbf{env}^*(\{\}).(x \circ e) \rightsquigarrow \lambda e{:}\mathbf{env}^*(\{\}).x \circ \big(e:\mathbf{env}^*(\{\}) \Rightarrow \mathbf{env}^*(\{x{:}*\})\big) : * \to *$
(CLam)

$\{x{:}*\} \vdash (x \circ e) \rightsquigarrow x \circ (e:\mathbf{env}^*(\{\}) \Rightarrow \mathbf{env}(\{x{:}*\})) : *$
(CCompA)

$\{e{:}\mathbf{env}^*(\{\})\} \vdash e \rightsquigarrow e : \mathbf{env}^*(\{\})$

$\Gamma = \{x{:}*\}\{e{:}*\}$

$\Gamma \vdash x \rightsquigarrow x : *$

$\vdash (1/x) \cdot id \rightsquigarrow (1/x) \cdot id : \mathbf{env}(\{x{:}\mathbf{int}\})$
(CExtn)

$\vdash 1 \rightsquigarrow 1 : \mathbf{int}$

$\vdash id \rightsquigarrow id : \mathbf{env}(\{\})$

The dynamic semantics of GTEC is provided by that of CIEC through the cast insertion, which will be defined in the following section.

Definition 1.5 (Var and DynEnvType) A set $\mathrm{Var}(M)$ of variables is variables appearing in a term M. $\mathrm{DynEnvType}(\{x_1,\ldots,x_n\})$ is a type $\{x_1:*\}\cdots\{x_n:*\}$.

3 CAST INTERNAL ENVIRONMENT CALCULUS

In this section, we propose a Cast Internal Environment Calculus, CIEC.

Definition 1.6 (Type of CIEC) *Types of CIEC are the same as the ones of GTEC, that is,*

$$B ::= \mathbf{int} \,|\, \mathbf{bool}$$

$$G ::= B \,|\, (* \to *) \,|\, \mathbf{env}\,(\Gamma)$$

$$T ::= B \,|\, (T \to T) \,|\, \mathbf{env}\,(\Gamma) \,|\, \mathbf{env}^*(\Gamma) \,|\, *$$

$$\Gamma ::= \{x_1:T_1\}\cdots\{x_n:T_n\}\ \text{where } n \geq 0$$

Definition 1.7 (Term of CIEC) *Terms of CIEC are defined by the following grammar.*

$$C ::= M \,|\, C:T_1 \Rightarrow T_2 \,|\, \mathbf{err}$$

where M is a term of GTEC.

Definition 1.8 (Typing of CIEC) A typing judgement $\Gamma\vdash_c M : T$ is a ternary relation among a type assignment Γ, a term M, and a type is defined inductively by the following rules.

$$\frac{\{x : T\} \in \Gamma}{\Gamma\vdash_c x : T}\,\text{Var} \quad \frac{\{c : T\} \in \Delta}{\Gamma\vdash_c c : T}\,\text{Const} \quad \frac{}{\Gamma\vdash_c \mathbf{err} : T}\,\text{Err}$$

$$\frac{\{x : S\}\Gamma\vdash_c M : T}{\{x : C\}\Gamma\vdash_c (\lambda x : S.M) : S \to T}\,\text{Lam} \quad \frac{\Gamma\vdash_c M : S \to T\,\Gamma\vdash_c N : S}{\Gamma\vdash_c (MN) : T}\,\text{App]}$$

$$\frac{}{\Gamma\vdash_c id : \mathbf{env}\,(\Gamma)}\,\text{Id} \quad \frac{\Gamma\vdash_c N : \mathbf{env}\,(\Gamma')\Gamma'\vdash_c M : T}{\Gamma\vdash_c (M \circ N) : T}\,\text{Comp}$$

$$\frac{\Gamma\vdash_c N : \mathbf{env}^*(\Gamma')\Gamma'' = \mathrm{Dyn}\,\mathrm{EnvType}\,(\mathrm{Var}\,(M)) \sqcup \Gamma'\Gamma'' \vdash M : T}{\Gamma\vdash_c (M \circ N) : T}\,\text{CompA}$$

$$\frac{\Gamma\vdash_c M : T\,\Gamma\vdash_c N : \mathbf{env}\,(\{x : S\}\Gamma')}{\Gamma\vdash_c (M/x) \cdot N : \mathbf{env}\,(\{x : T\}\Gamma')}\,\text{Ext} \quad \frac{\Gamma\vdash_c M : T\,\Gamma\vdash_c N : \mathbf{env}^*(\{x : S\}\Gamma')}{\Gamma\vdash_c (M/x) \cdot N : \mathbf{env}^*(\{x : T\}\Gamma')}\,\text{ExtA}$$

$$\frac{\Gamma\vdash_c C : S\,S \sim T}{\Gamma\vdash_c (C : S \Rightarrow T) : T}\,\text{Cast}$$

Example 1.2 (Type Derivation of CIEC) We show an example of derivation of typing in CIEC as follows.

$$\vdash_c \left(\lambda e{:}*.x \circ (e : \mathbf{env}^*(\{\})\Rightarrow\mathbf{env}(\{x{:}*\}))\right)\left((1/x) \cdot id : \mathbf{env}(\{x{:}\mathbf{int}\})\Rightarrow*\right) : *$$

- $\vdash_c \lambda e{:}*.x \circ (e : \mathbf{env}^*(\{\})\Rightarrow\mathbf{env}(\{x{:}*\})) : * \to *$
  - $\{e{:}*\} \vdash_c x \circ (e : \mathbf{env}^*(\{\})\Rightarrow\mathbf{env}(\{x{:}*\})) : *$
    - $\{e{:}*\} \vdash_c e : \mathbf{env}^*(\{\})\Rightarrow\mathbf{env}(\{x{:}*\}) : \mathbf{env}(\{x{:}*\}\{e{:}*\})$
      - $\{x{:}*\} \vdash_c e : \mathbf{env}^*(\{\})$
      - $\mathbf{env}^*(\{\}) \sim \mathbf{env}(\{x{:}*\}\{e{:}*\})$
- $\vdash_c (1/x) \cdot id : \mathbf{env}(\{x{:}\mathbf{int}\})\Rightarrow* : *$
  - $\vdash_c (1/x) \cdot id : \mathbf{env}(\{x{:}\mathbf{int}\})$
    - $\vdash_c 1 : \mathbf{int}$
    - $\vdash_c id : \mathbf{env}(\{\})$
  - $\mathbf{env}(\{x{:}\mathbf{int}\}) \sim *$

Definition 1.9 (Value of CIEC) A set of values v and a set of environment values E_v of CIEC are subsets of its terms are defined by the following grammar.

$$v ::= c|\lambda x : T.C|(\lambda x : T.C) \circ E_v|v : T_1 \to T_2 \Rightarrow T_3 \to T_4|v : G \Rightarrow \mathbf{env}^*(\Gamma)|E_v$$

$$E_v ::= id|(v/x) \cdot E_v$$

Definition 1.10 (Evaluation Contexts of CIEC) Evaluation contexts of CIEC are defined by the follwoing grammar.

$$E ::= [\,]|(E[\,]C)|(vE[\,])|(C \circ E[\,])|(E[\,]/x) \cdot C|(v/x) \cdot E[\,]|E[\,] : S \Rightarrow T$$

Definition 1.11 (Consistency on Environment Types) We define a consistency $\mathbf{env}\,(\Gamma)\approx \mathbf{env}\,(\Gamma')$ on environment types $\mathbf{env}\,(\Gamma)$ and $\mathbf{env}\,(\Gamma')$ as

- codomain $(\Gamma) = $ codomain (Γ'),
- $T \sim T'$ for $\{x : T\} \in \Gamma$ and $\{x : T'\} \in \Gamma$.

A set codomain of variables is the codomain of Γ if you regard Γ as a partial function of variables to types.

Definition 1.12 (Reduction Rules of CIEC) A reduction $M \to M'$ is a binary relation between terms of CIEC defined inductively by the following rules.

$$v \circ id \to v \quad \text{(IdL)}$$

$$id \circ v \to v \quad \text{(IdR)}$$

$$((\lambda x : S.M \circ E_v)v) \to C \circ ((v/x) \cdot E_v) \quad \text{(Beta1)}$$

$$((\lambda x : S.M)v) \to C \circ ((v/x) \cdot id) \quad \text{(Beta2)}$$

$$x \circ ((v/x) \cdot E_v) \to v \quad \text{(VarRef)}$$

$$y \circ ((v/x) \cdot E_v) \to y \circ E_v \quad \text{(VarSkip)}$$

$$((C_1/x) \cdot C_2) \circ E_v \to ((C_1 \circ E_v)/x) \cdot (C_2 \circ E_v) \quad \text{(DExtn)}$$

$$(C_1 C_2) \circ E_v \to (C_1 \circ E_v)(C_2 \circ E_v) \quad \text{(DApp)}$$

$$(C : S \Rightarrow T) \circ E_v \to (C \circ E_v) : S \Rightarrow T \quad \text{(DCast)}$$

$$(C_1 \circ C_2) \circ E_v \to C_1 \circ (C_2 \circ E_v) \quad \text{(Assoc)}$$

$$E[C] \to E[C'] \ \textit{if} \ C \to C' \quad \text{(Cong)}$$

$$E[\mathbf{err}] \to \mathbf{err} \ \textit{if} \ E[\ ] \neq [\ ] \quad \text{(ErrCong)}$$

$$\mathbf{err} \circ E_v \to \mathbf{err} \quad \text{(ErrComp)}$$

$$C \circ v \to \mathbf{err} \ \textit{if} \ v \ \textit{is not} \ E_v \quad \text{(ErrEval)}$$

$$(v_1/x) \cdot v_2 \to \mathbf{err} \ \textit{if} \ v_2 \ \textit{is not} \ E_v \quad \text{(ErrExtn)}$$

$$v : B \Rightarrow B \to v \quad \text{(IdBase)}$$

$$v : * \Rightarrow * \to v \quad \text{(IdStar)}$$

$$v : G \Rightarrow * \Rightarrow G \to v \quad \text{(Succeed)}$$

$$v : G_1 \Rightarrow * \Rightarrow G_2 \to \mathbf{err} \ \textit{if} \ G_1 \neq G_2 \quad \text{(Fail)}$$

$$(v_1 : T_1 \to T_2 \Rightarrow T_3 \to T_4)v_2 \to (v_1(v_2 : T_3 \Rightarrow T_1)) : T_2 \Rightarrow T_4 \quad \text{(AppCast)}$$

$$v : T \Rightarrow env^*(\Gamma) \to v : T \Rightarrow G \Rightarrow env^*(\Gamma)$$

$$\textit{if} \ T \neq env^*(\Gamma), T \neq G, T \sim G \quad \text{(Ground)}$$

$$v : \mathbf{env}^*(\Gamma) \Rightarrow T \to v : \mathbf{env}^*(\Gamma) \Rightarrow G \Rightarrow T$$

$$\text{if } T \neq \mathbf{env}^*(\Gamma), T \neq G, T \sim G \text{ (Expand)}$$

$$\overline{(v_n/x_n) \cdot id} : \mathbf{env}(\Gamma) \Rightarrow \mathbf{env}^*(\Gamma'') \Rightarrow \mathbf{env}^*(\Gamma')$$

$$\rightarrow \overline{(v_n : T_n \Rightarrow T_{n'}/x_n) \cdot id}$$

$$\text{if } \mathbf{env}(\Gamma) \approx \mathbf{env}(\Gamma') \text{ (ESucceed)}$$

$$\rightarrow err$$

$$\text{if } env(\Gamma) env(\Gamma') \text{ (EFail)} .$$

$$(v_1/x) \cdot v_2 : T' \Rightarrow \mathbf{env}^*(\Gamma'')$$
$$: \mathbf{env}^*(\{x : T\}\Gamma') \Rightarrow \mathbf{env}^*(\{x : T\}\Gamma')$$

$$\rightarrow (v_1/x) \cdot v_2 : \mathbf{env}(\{x : T\}\Gamma') \Rightarrow \mathbf{env}^*(\{x : T\}\Gamma')$$

$$\text{if } T' = \mathbf{env}(\Gamma) \text{ (CastOut)}$$

$$\rightarrow err$$

$$\text{if } T' \neq \mathbf{env}(\Gamma) \text{ (ECastOut)}$$

4 CAST INSERTION AND TYPE SAFETY

Definition 1.13 (Type Precision $\sqsubseteq$) Type precision $S \sqsubseteq T$ is a binary relation between types defined by the following rules.

$$T \sqsubseteq *B \sqsubseteq B \frac{T_1 \sqsubseteq T_3 \ T_2 \sqsubseteq T_4}{(T_1 \rightarrow T_2) \sqsubseteq (T_3 \rightarrow T_4)}$$

Definition 1.14 (Operator $\sqcup$)

$$(\{x_1 : S_1\} \cdots \{x_n : S_n\}\{y_1 : T_1\} \cdots \{y_l : T_l\}) \sqcup (\{x_1 : S_{1'}\} \cdots \{x_n : S_{n'}\}\{z_1 : U_1\} \cdots \{z_m : U_m\})$$

$$= \{x_1 : S_{1''}\} \cdots \{x_n : S_{n''}\}\{y_1 : T_1\} \cdots \{y_l : T_l\}\{z_1 : U_1\} \cdots \{z_m : U_m\}$$

where
- $S_{i''} = S_i$ if $S_i \sqsubseteq S_{i'}$,
- $S_{i''} = S_{i'}$ if $S_{i'} \sqsubseteq S_i$, and
- $S_{i''} = S_i$, otherwise.

Definition 1.15 (Cast-insertion Rule of *GTEC*) *Cast insertion $\Gamma \vdash MC : T$ is a quaternary relation among a type assingment Γ, a term M, a term C, and a type T.*

$$\frac{\{x{:}T\} \in \Gamma}{\Gamma \vdash x \leadsto x : T} \text{ CVar} \quad \frac{\{c{:}T\} \in \Delta}{\Gamma \vdash c \leadsto c : T} \text{ CConst} \quad \frac{\{x{:}S\}\Gamma \vdash M \leadsto C : T}{\Gamma \vdash \lambda x{:}S.M \leadsto \lambda x{:}S.C : S \to T} \text{ CLam}$$

$$\frac{\Gamma \vdash M \leadsto C_1 : T_1 \to T_2 \quad \Gamma \vdash N \leadsto C_2 : T_1' \quad T_1' \sim T_1}{\Gamma \vdash (MN) \leadsto C_1(C_2 : T_1' {\Rightarrow} T_1) : T_2} \text{ CApp}$$

$$\frac{}{\Gamma \vdash (M\ N) \leadsto (C_1 : *{\Rightarrow}* \to *{\Rightarrow})(C_2 : T{\Rightarrow}*) : *} \text{ CAppA}$$

$$\frac{}{\Gamma \vdash id \leadsto id : \mathbf{env}(\Gamma)} \text{ CId} \quad \frac{\Gamma \vdash N \leadsto C_2 : \mathbf{env}(\Gamma') \quad \Gamma' \vdash M \leadsto C_1 : T_1}{\Gamma \vdash (M \circ N) \leadsto (C_1 \circ C_2) : T_1} \text{ CComp}$$

$$\frac{\Gamma \vdash N \leadsto C_2 : \mathbf{env}^*(\Gamma') \quad \Gamma'' = \mathrm{DynEnvType}(\mathrm{Var}(M)) \sqcup \Gamma' \quad \Gamma'' \vdash M \leadsto C_1 : T_1}{\Gamma \vdash (M \circ N) \leadsto C_1 \circ (C_2 : \mathbf{env}^*(\Gamma') {\Rightarrow} \mathbf{env}^*(\Gamma'')) : T_2} \text{ CCompA}$$

$$\frac{\Gamma \vdash M \leadsto C_1 : T_1 \quad \Gamma \vdash N \leadsto C_2 : \mathbf{env}(\{x{:}S\}\Gamma')}{\Gamma \vdash (M/x) \cdot N \leadsto (C_1/x) \cdot C_2 : \mathbf{env}(\{x{:}T_1\}\Gamma')} \text{ CExtn}$$

$$\frac{\Gamma \vdash M \leadsto C_1 : T_1 \quad \Gamma \vdash N \leadsto C_2 : \mathbf{env}^*(\{x{:}S\}\Gamma')}{\Gamma \vdash (M/x) \cdot N \leadsto (C_1/x) \cdot C_2 : \mathbf{env}^*(\{x{:}T_1\}\Gamma')} \text{ CExtnA}$$

4.1   Type safety

Type safety of GTEC is that, if a typed term of GTEC is given, then its cast inserted term is evaluated to either a value or a type mismach is found by the inserted casts.

Theorem 1 *For a term M of GTEC, a type assignment Γ, and a type T, suppose $\Gamma \vdash M : T$ and $\Gamma \vdash MC : T$, then one of the following conditions holds:*

- C is reduced to a value of type T; $C \to v$ and $\Gamma \vdash v : T$,
- $C \to \mathbf{err}$,
- $C \to C'$ and $\Gamma \vdash C' : T$, where C is neither a value nor $\mathbf{err}$,
- $T = \mathbf{env}(\Gamma')$, $C \to C'$, $\Gamma \vdash C' : \mathbf{env}(\Gamma'')$, and $\mathbf{env}(\Gamma') \approx \mathbf{env}(\Gamma'')$.

In other words, if $C \to \cdots \to C''$, then C'' is either a value or $\mathbf{err}$, keeping the typing.
This theorem is derived from the following threee lemmas.

Lemma 1.1 If $\Gamma \vdash M : T$ then $\Gamma \vdash MC : T$ and $\Gamma \vdash C : T$.

Proof. This lemma is proved by induction on the structure of $\Gamma \vdash M : T$. We show some selected cases.

Case of AppA. Suppose that

$$\frac{\Gamma \vdash M_1 : * \Gamma \vdash M_2 : T}{\Gamma \vdash (M_1 M_2) : *} \text{ AppA}$$

By the induction hypothesis, we have that $\Gamma \vdash M_1 C_1 : *$, $\Gamma \vdash_c C_1 : *$, $\Gamma \vdash M_2 C_2 : T$, $\Gamma \vdash_c C_2 : T$. By rule (CAppA), we know $\Gamma \vdash (M_1 M_2)(C_1 : * \Rightarrow * \to *)(C_2 : T \Rightarrow *) : *$ Since $* \sim (* \to *)$ and $* \sim T$, it is derived that $\Gamma \vdash_c (C_1 : * \Rightarrow * \to *) : * \to *$ and $\Gamma \vdash_c (C_2 : T \Rightarrow *) : *$. Then, by rule (App), we know $\Gamma \vdash_c (C_1 : * \Rightarrow * \to *)(C_2 : T \Rightarrow *) : *$.

Case of ExtA. Suppose that

$$\frac{\Gamma \vdash M_1 : T \Gamma \vdash M_2 : \{x : S\} \, \mathbf{env}^*(\Gamma')}{\Gamma \vdash (M_1/x) \cdot M_2 : \, \mathbf{env}^*(\{x : T\}\Gamma')} \mathrm{ExtA}$$

for some type S.

By the induction hypothesis, we have that $\Gamma \vdash M_2 C_2 : T$, $\Gamma \vdash_c C_1 : T$, $\Gamma \vdash M_2 C_2 : env^*(\{x : S\}\Gamma')$, and $\Gamma \vdash_c C_2 : \mathbf{env}^*(\Gamma')$.

By rule CExtA,

$$\Gamma \vdash (M_1/x) \cdot M_2(C_1/x) \cdot C_2 : \, env^*(\{x : T\}\Gamma').$$

By rule ExtA,

$$\Gamma \vdash_c (C_1/x) \cdot C_2 : \, env^*(\{x : T\}\Gamma').$$

Q.E.D.

Lemma 1.2 If $\Gamma \vdash_c C : T$ and $C \to C'$, then it holds that

• $\Gamma \vdash_c C' : T$,
• $T = \mathbf{env}(\Gamma')$, $C \to C'$, $\Gamma \vdash_c C' : \mathbf{env}(\Gamma'')$, $\mathbf{env}(\Gamma') \approx \mathbf{env}(\Gamma'')$. for some Γ' and Γ'', or
• $C' = \mathbf{err}$.

holds.

Proof. This lemma is proved by induction on the structure of $\Gamma \vdash_c C : T$. We show some selected cases.

Case of Ext. Suppose that

$$\frac{\Gamma \vdash_c C_1 : T_1 \Gamma \vdash_c C_2 : \mathbf{env}(\{x : S\}\Gamma')}{\Gamma \vdash_c (C_1/x) \cdot C_2 : \{x : T_1\}\Gamma'} \mathrm{Ext}$$

The reduction $C \to C'$ can be derived by rule Cong or ErrExtn .

Subcase of Cong . It holds either

• $C_2 \to C_2'$ for some C_1', or
• $C_2 \to C_2'$ for some $C_{2'}$.

In the former case, we have $\Gamma \vdash_c C_1' : T_1$ by the induction hyposesis, and therefore, $\Gamma \vdash_c (C_1'/x) \cdot C_2 : \{x : T_1\}\Gamma'$. In the latter case, we have $\Gamma \vdash_c C_2' : \mathbf{env}(\Gamma')$ by the induction hypothesis, and therefore, $\Gamma \vdash_c (C_1'/x) \cdot C_{2'} : \{x : T_1\}\Gamma'$.

Q.E.D.

Lemma 1.3 If $\Gamma \vdash C : T$, then one of the folloings holds:

• C is a value;
• C is err;
• $C \to C'$ for some C'.

This lemma is proved by induction on $\Gamma \vdash C : T$ straight-forwardly.

4.2 *Concluding remarks*

In this paper, we proposed two computational systems: the Gradually Typed Environment Calculus (GTEC) and the Cast Internal Environment Calculus (CIEC). The former system models a programming language in which type conversion is made implicitly and the latter a programming language in which type conversion is provided by using cast operator. Then, we introduce a translation of the GTEC into the CIEC, which inserts casts into an expression appropriately.

In the previous work, we proposed a simply-typed lambda calculus with first-class environments and continuations (Nishizaki 2018). One of the future research direction is an extension of GTEC and CIEC to such a calculus.

REFERENCES

Abadi, M., L. Cardelli, P.-L. Curien, & J.-J. Lévy (1991, October). Explicit substitutions. *Journal of Functional Programming 1*(4), 375–416.

Girard, J.-Y., P. Taylor, & Y. Lafont (1989). *Proofs and Types*, Volume 7 of *Cambridge Tracts in Computer Science*. Cambridge University Press.

Igarashi, Y., T. Sekiyama, & A. Igarashi (2017, August). On polymorphic gradual typing. *Proc. ACM Program. Lang. 1*(ICFP), 40:1–40:29.

Nishizaki, S. (1994). ML with first-class environments and its type inference algorith. In *Lecture Notes in Computer Science*, Volume 792, pp. 95–116. Springer-Verlag Berlin Heidelberg.

Nishizaki, S. (1995). Simply typed lambda calculus with first-class environments. *Publications of Research Institute for Mathematical Sciences Kyoto University 30*(6), 1055–1121.

Nishizaki, S. (2018). Untyped call-by-value calculus with first-class continuations and environments. Accepted.

Nishizaki, S. & M. Fujii (2012). Strong reduction for typed lambda calculus with first-class environments. In *Lecture Notes in Computer Science*, Volume 7473, pp. 632–639. Springer-Verlag Berlin Heidelberg.

Siek, J. G. & W. Taha (2006). Gradual typing for functional languages. In *Scheme and Functional Programming Workshop*, pp. 81–92.

Siek, J. G., M. M. Vitousek, M. Cimini, & J. T. Boyland (2015). Refined Criteria for Gradual Typing. In *1st Summit on Advances in Programming* Languages, *SNAPL 2015*, pp. 274–293.

APPENDIX

For convenience, we will introduce the notions related to the gradually Typed Lambda Calculus, the Cast Internal Calculus, and the cast insertion, introduced in Paper (Siek, Vitousek, Cimini, & Boyland 2015) (Siek, Vitousek, Cimini, & Boyland 2015)..

Gradually Typed Lambda Calculus

Definition 1.16 (Term of GTLC) Terms of GTLC are defined by the following grammar:

$$M ::= c \mid x \mid \lambda x : T.M \mid (MN)$$

Definition 1.17 (Type of GTLC) Basic types are defined as an integer type **int** or a boolean type **bool**:

$$B ::= \mathbf{int} \mid \mathbf{bool}$$

Types of GTLC are defined by the following grammar:

$$T ::= B \mid (T \to T) \mid *$$

Definition 1.18 (Type Consistency of GTLC) *Type consistency is a binary relation defined by the following rules:*

$$\frac{}{* \sim T} \quad \frac{}{T \sim *} \quad \frac{}{B \sim B} \quad \frac{T_1 \sim T_3 \quad T_2 \sim T_4}{(T_1 \to T_2) \sim (T_3 \to T_4)}$$

Definition 1.19 (Typing Rules of GTLC) Typing rules of GTLC are as follows:

$$\frac{x : T \in \Gamma}{\Gamma \vdash x : T}\text{Var} \quad \frac{c : T \in \Delta}{\Gamma \vdash c : T}\text{Const} \quad \frac{\{x : S\}\Gamma \vdash M : T}{\Gamma \vdash \lambda x : S.M : (S \to T)}\text{Lam}$$

$$\frac{\Gamma \vdash M : T_1 \; \Gamma \vdash N : T_2 \; \text{fun}(T_1) = (T_{11} \to T_{12}) \; T_2 \sim T_{11}}{\Gamma \vdash (MN) : T_{12}}\text{App}$$

A constant type assignment Δ is a correspondence between constants and their types, which consists of pairs of a constant and its type.

Definition 1.20 (fun) A function fun is a partial function between types is defined as follows:

$$\text{fun}(T_1 \to T_2) = (T_1 \to T_2)$$

$$\text{fun}(*) = (* \to *)$$

$$\text{fun}(B) = \textbf{undefined}$$

Cast Inserted Lambda Calculus

Definition 1.21 (Types of CILC) A basic type of CILC is either the integer type **int** or **bool**:

$$B ::= \textbf{int} \,|\, \textbf{bool}.$$

A ground type of CILC is defined as follows:

$$G ::= B|(* \to *).$$

A type of CILC is defined by the following grammar:

$$T ::= B|(T \to T)|*$$

A type assignment Γ is a partial mapping whose domain is a finite set of variables and codomain is a finite set of types, which can be also defined by the following grammar:

$$\Gamma ::= \{\}|\{x : T\}\Gamma.$$

Definition 1.22 (Terms of CILC) A set of the terms of CILC is defined as a superset of the terms of GTLC by the following grammer:

$$C ::= M|C : T_1 \Rightarrow T_2|\textbf{err}$$

Definition 1.23 (Typing rules of CILC) Typing rules of CILC consists of the typing rules of GTLC and the following rule:

$$\frac{\Gamma \vdash C : T_1 \; T_1 \sim T_2}{\Gamma \vdash (C : T_1 \Rightarrow T_2) : T_2}\text{Cast}$$

Definition 1.24 (Cast-insertion Rule) Cast insertion $\Gamma \vdash MC : T$ is a quaternary relation among a type assingment Γ, a term M of GTLC, a term C of CITL, and a type T.

$$\frac{\{x{:}T\} \in \Gamma}{\Gamma \vdash x \leadsto x : T} \; \text{CVar} \quad \frac{\{c{:}T\} \in \Delta}{\Gamma \vdash c \leadsto c : T} \; \text{CConst} \quad \frac{\{x{:}S\}\Gamma \vdash M \leadsto C : T}{\Gamma \vdash \lambda x{:}S.M \leadsto \lambda x{:}S.C : S \to T} \; \text{CLam}$$

$$\frac{\Gamma \vdash M \leadsto C : T_1 \quad \Gamma \vdash N \leadsto C' : T_2 \quad \text{fun}(T_1) = (T_{11} \to T_{12}) \quad T_2 \sim T_{11}}{\Gamma \vdash (MN) \leadsto C : T_1 \Rightarrow ((T_{11} \to T_{12})(C' : T_2 \Rightarrow T_{11})) : T_{12}} \; \text{CApp}$$

$\vdash (\lambda f{:}*.f 2)(\lambda x{:}\mathbf{int}.x) \leadsto$
$\quad (\lambda f{:}*.(f : * \Rightarrow * \to *)(2 : \mathbf{int} \Rightarrow *)) (\lambda x{:}\mathbf{int}.x : \mathbf{int} \to \mathbf{int} \Rightarrow *)$
$\quad : *$
(CApp)
$\vdash$—fun$(* \to *) = (* \to *)$
$\vdash$—$* \sim (* \to *)$
$\vdash$— $\quad \vdash \lambda f{:}*.(f\,2) \leadsto \lambda f{:}*.(f : * \Rightarrow * \to *)(2 : \mathbf{int} \Rightarrow *) : * \to *$
$\qquad$ (CLam)
$\qquad \quad \llcorner \;\; \{f{:}*\} \vdash (f\,2 \leadsto (f : * \Rightarrow * \to *)(2 : \mathbf{int} \Rightarrow *) : * \to *$
$\qquad\qquad$ CApp
$\qquad\qquad$— fun$(*) = (* \to *)$
$\qquad\qquad$— $* \sim \mathbf{int}$
$\qquad\qquad$— $\{f{:}*\} \vdash f \leadsto f : *$
$\qquad\qquad$ (CVar)
$\qquad\qquad\quad \llcorner \{f{:}*\} \in \{f{:}*\}$
$\qquad\qquad$— $\{f{:}*\} \vdash 2 \leadsto 2 : \mathbf{int}$
$\qquad\qquad$ (CConst)
$\qquad\qquad\quad \llcorner \{2{:}\mathbf{int}\} \in \Delta$
$\vdash$— $\vdash \lambda x{:}\mathbf{int}.x \leadsto \lambda x{:}\mathbf{int}.x : \mathbf{int} \to \mathbf{int}$
$\qquad$ (CLam)
$\qquad \quad \llcorner \;\; \{x{:}\mathbf{int}\} \vdash x \leadsto x : \mathbf{int}$
$\qquad\qquad$ (CVar)
$\qquad\qquad\quad \llcorner \{x{:}\mathbf{int}\} \in \{x{:}\mathbf{int}\}$

Definition 1.25 (Value of CILC) A value v of CILC is a term satisfying the following grammar:

$$v ::= c \mid \lambda x : T.C \mid v : T_1 \to T_2 \Rightarrow T_3 \to T_4 \mid v : G \Rightarrow *$$

Definition 1.26 (Evaluation Context of CILC) Evaluation contexts of CILC are defined by the folllowing grammar:

$$E[\,] ::= c \mid (E[\,]C) \mid (vE[\,]) \mid E[\,] : T_1 \Rightarrow T_2$$

Definition 1.27 *Reduction Rule of CILC A reduction relation $(-) \to (-)$ is defined by the following reduction rules.*

$$(\lambda x : T.C)v \to C[x := v]$$

$$v : B \Rightarrow B \to v$$

$$v : * \Rightarrow * \to v$$

174

$v : G \Rightarrow * \Rightarrow G \to v$

$v : G_1 \Rightarrow * \Rightarrow G_2 \to err(G_1 \neq G_2)$

$(v_1 : T_1 \to T_2 \Rightarrow T_3 \to T_4)v_1 \to (v_1(v_2 : T_3 \Rightarrow T_1)) : T_2 \Rightarrow T_4$

$v : T \Rightarrow * \to v : T \Rightarrow G \Rightarrow *(T \neq *, T \neq G, T \sim G)$

$v : * \Rightarrow T \to v : * \Rightarrow G \Rightarrow T(T \neq *, T \neq G, T \sim G)$

$E[C] \to E[C'] \, if \, C \to C'$

$E[\mathbf{err}] \to \mathbf{err}$

Theory and Practice of Computation – Nishizaki et al (eds)
© 2021 Taylor & Francis Group, London, ISBN 978-0-367-41473-3

A directed minimum connected dominating set for protein-protein interaction networks

Y.L. Briones
Department of Chemistry, Ateneo de Manila University, Quezon City, NCR, Philippines
Philippine Genome Center, University of the Philippines Diliman, Quezon City, NCR, Philippines

M.R. Castro & R.D. Jalandoni
National Institute of Physics, University of the Philippines Diliman, Quezon City, NCR, Philippines
Philippine Genome Center, University of the Philippines Diliman, Quezon City, NCR, Philippines

H.N. Adorna
Department of Computer Science, University of the Philippines Diliman, Quezon City, NCR, Philippines

J.A. Dizon & A.T. Young
Philippine Genome Center, University of the Philippines Diliman, Quezon City, NCR, Philippines

ABSTRACT: Network analysis of Protein-Protein Interaction (PPI) maps is valuable for identifying key regulators of cellular processes such as cancer-related signaling pathways. Previous studies have used the Minimum Connected Dominating Set (MCDS) to identify such key proteins. However, these studies did not account for directionality and regulatory effects. In this study, a directed MCDS algorithm was developed and tested on a human PPI network simulating HER2-positive breast cancer using regulation and expression data. The biological significance of the directed MCDS was examined using pathway enrichment analysis. The directed MCDS was found to be significantly smaller (333 nodes) than the undirected MCDS (811 nodes) of the same network, with 68 uniquely identified nodes involved in cancer. This suggests that the directed MCDS identifies a more specialized set of critical proteins under given biological conditions. The performance of the directed MCDS under conditions other than breast cancer is recommended for future analysis.

Keywords: Protein-protein interactions, minimum connected dominating set, breast cancer, network analysis, signal transduction

1 INTRODUCTION

Protein-protein interactions (PPIs) control signal transduction pathways and other biological processes that occur within the cell. A PPI network can be perturbed by significant changes in expression of a protein, causing diseases such as cancer (Goncearenco et al. 2017). One example is the overexpression of the human epidermal growth factor receptor 2 protein (HER2 or ERBB2), which is a diagnostic biomarker for certain types of breast cancer (Mnard et al. 2000). Many studies have explored PPI networks to find master regulator proteins which may serve as effective drug targets (Goncearenco et al. 2017; Amala & Emerson, 2019; Ivanov et al. 2013; Nazarieh et al. 2016; Skwarczynska & Ottmann, 2015). PPI networks can be mathematically analyzed using graph theory as networks with proteins as nodes and interactions as edges (Fionda, 2019). According to Bhowmick & Seah (2014), when studying dense PPI networks, nodes can be grouped according to two modules: the topological and functional modules. The topological module refers to the clustering of nodes according to their nearest neighbors. Meanwhile, the functional module refers to the clustering of nodes according to biological function. Studies have shown that topologically and biologically significant nodes commonly overlap (Barabsi et al. 2010; Kashyap et al. 2018).

Thus, computational methods that identify topological hubs are valuable for biological analysis of PPI networks. One such method is finding the minimum dominating set (MDS), which identifies hubs but not the connections between them. A less explored but promising method is finding the minimum connected dominating set (MCDS), which identifies hubs and their connections to each other, forming a backbone of the network. The connectivity of the MCDS allows for analysis of not only hub proteins but their interactions and roles in pathways. However, existing studies on the MCDS of PPI networks such as that of Dizon & Malolos (2018) do not account for important biological factors such as directionality or regulation. This may bias results towards topological significance over functional significance.

To address this issue, we developed a heuristic MCDS algorithm that accounts for directionality and regulation between proteins in the network. By modeling our algorithm on biological principles, we expected to obtain results with higher functional significance. In order to test this hypothesis, we used the SIGNOR database of regulated protein networks for our study (Lo Surdo et al. 2017). By modeling our algorithm on biological principles, we expected to obtain results with higher functional significance. We tested the performance of the directed MCDS in a PPI network simulating HER2-positive breast cancer and compared it to the undirected MCDS. Ultimately, the study aims to move towards a more dynamic approach to PPI network analysis.

Limitations of the study include the incompleteness of the SIGNOR database, as it consists of known causal relationships based on regulation data. The study also only looks at one biological condition (HER2 + breast cancer). The directed MCDS must be run multiple times and evaluated under different biological conditions to comprehensively test its performance.

2 REVIEW OF RELATED LITERATURE

2.1 *Computational analysis of gene regulatory networks*

Computational techniques have proven useful in pruning large gene regulatory networks (GRNs) to their most critical players. Gene regulatory networks are directed graphs where nodes represent genes and edges represent regulatory relationships. In a study by Nazarieh et al. (2016), the MDS and MCDS of various GRNs successfully identified key player gene products in the networks. The study found that the MDS best identified driver genes or dominating hubs of the networks. Meanwhile, the MCDS (using a heuristic algorithm) identified experimentally validated master regulators which control other genes in the network. Master regulators are promising drug targets since they have a cascade effect on the network. The potential of a heuristic MCDS to identify master regulators in a directed graph is a valuable finding that has yet to be explored in protein-protein interaction networks.

2.2 *Heuristic MCDS of an undirected PPI network*

A heuristic MCDS algorithm was applied to an undirected PPI network of human proteins (from the HINT database) in a study by Dizon & Malolos (2019). The HINT database is a large collection of eight different human PPI databases (Das & Yu, 2012). Thus, the MCDS generated was large (1,551 nodes). Enrichment analysis using Gene Ontology (GO) found that proteins in the MCDS were important to the framework of the human PPI network. The results of the MCDS were compared against the centrality-connected minimum dominating set (CC-MDS) using the algorithm of Zhang et al. (2015). They found that the MCDS contained more unique nodes than the CC-MDS.

These unique proteins are involved in essential cellular processes such as cell growth, cell reproduction and protein folding. The findings illustrate the potential of the MCDS to identify critical proteins and provides a starting point for our study.

2.3 *HER2 overexpression and signaling in breast cancer*

The analysis of PPI networks from an oncological perspective can offer valuable insights on cancer signaling pathways and possible treatment methods. The mechanisms of HER2 signal transduction in breast cancer has been extensively studied, making it an ideal model case for biological analysis of the directed MCDS.

A review by Moasser (2007) details the role of HER2 in breast cancer. The HER2 protein belongs to the epidermal growth factor receptor (EGFR) family of transmembrane receptors that respond to extracellular signals by stimulating intracellular signal transduction activity. The overexpression of HER2 is known to induce neoplastic (does not cause metastasis, i.e., malignancy) cell transformation and serves as an important biomarker for HER2-positive breast cancer-targeted drugs. Overexpressed HER2 forms heterodimers with HER3, cooperatively stimulating oncogenic signaling mainly through MAPK and PI3K-Akt signal transduction pathways. These pathways are known to induce proliferation, angiogenesis, and anti-apoptotic behavior, ultimately promoting breast tumorigenesis. HER2 overexpression has also been shown to play a role in altering immune signaling pathways (Nicolini et al. 2018) and trigger a pro-inflammatory response by interleukin signaling (Liu et al. 2018). Gene expression data on HER2+ breast cancer is widely available, making it possible to simulate and study the condition in a human PPI network.

Given that conditions such as cancer drastically change the state of a PPI network, network analysis must be dynamic rather than static. This calls for the need to develop a directed MCDS that can account for changes in regulation or expression so that the resulting set of key proteins is more specific to the given biological condition.

3 METHODOLOGY

3.1 *Minimum connected dominating set*

Let $G = (V, E)$ where V is the finite set of n nodes, and E is a finite set of edges be a directed graph. The dominating set of a network is the subset D of V, such that each node $v \varepsilon D$ is adjacent to all nodes in G that does not belong to D. By getting the minimum among all possible subset D of V, and restrict nodes in D to be connected, the minimum connected dominating set of nodes of a graph G is achieved.

3.2 *Maximum leaf spanning tree algorithm*

The MCDS problem is an NP-hard problem (Parthiban et al. 2015). Many studies suggested different algorithms in solving for the MCDS (Simonetti & Lucena, 2011; Tian & Ding, 2013; Purohit & Sharma, 2010). Dizon et al (2019) used the maximum leaf spanning tree (MLST) algorithm to determine the MCDS of an undirected network. We use in this work a modified version of the MLST algorithm to consider a directed graph.

3.3 *Algorithm*

The directed MCDS of the SIGNOR dataset was obtained using a code implemented in Python programming language. Figure 1 shows the flowchart of the algorithm. The code used for this study can be found in https://github.com/mcastro2/Trim_MCDS.

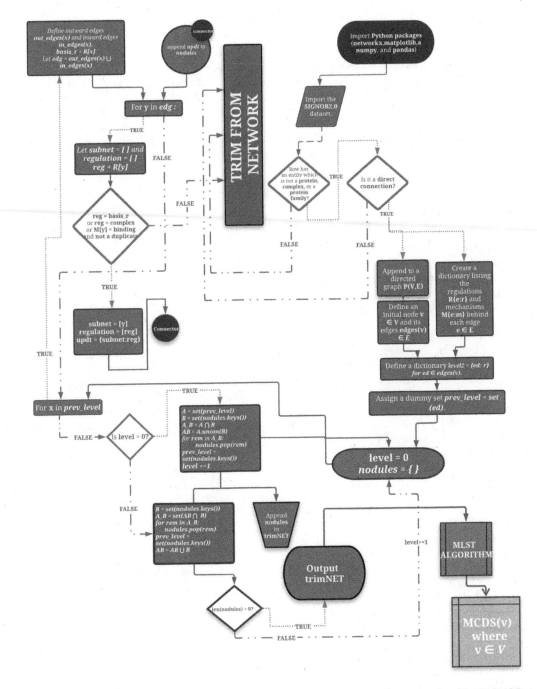

Figure 1. trimNET algorithm flowchart for parsing and building of a subnetwork from the SIGNOR database using Python. The algorithm utilizes a general classification of each of the interactions as either up-regulation, down-regulation, or complex forming interactions. Trimming of the initial network was performed on the basis of these classifications.

3.4 Parsing of SIGNOR 2.0 dataset

The Signal Network Open Resource (SIGNOR) database is a collection of more than 22000 manually-annotated causal relationships between proteins that participate in signal transduction (Lo Surdo et al. 2017). The database includes a pair of entities, their types, their database and database ID, their effect or type of regulation, the mechanism, and other information about each interaction. We limited the range of our analysis by parsing interactions that are only between proteins, complexes, and protein families. Parameters were also limited to name of entity, type of regulation, and their corresponding mechanisms. All indirect connections were then parsed out as to mitigate the number of disconnected proteins.

3.5 Minimizing the size of the parsed data via trimNET

The resulting parsed dataset contains 10698 data points for each interaction. Due to the great size of the network, it was trimmed down using the following algorithm:

trimNET Algorithm
Input: A directed graph $G = (V, E)$
Output: Dictionary of edges of the trimmed network *trimNET*
Algorithm:
Step 1: Initialize the algorithm by defining a blank, unordered key-value-pair list *trimNET* = {}

Step 2: Select an initial node $v \in V$ at which the trimmed network shall be centered. The selected node must represent a protein that is an important biomarker for breast cancer positive networks. Determine $edges(v) = edges_{inward} \cup edges_{outward}$ of the initial node. Note the regulation r for each of the connections. Afterwards, define an unordered key-value-pair list level2 where $v_{edge} \; \varepsilon \; edges(v)$ are keys with their corresponding regulations $r_i \; \varepsilon \; r$ as their values.

Step 3: Assign a dummy set $prev_{level}$ to contain the keys of the previous level such that $prev_{level} = (edges(v))$. Afterwards, copy the key-value-pairs of $level_2$ into the trimmed network trimNET such that $trimNET = v_{edge} : r_i$ where $v_{edge} \; \varepsilon \; edges(v)$ and $r_i \; \varepsilon \; r$.

Step 4: For each edge $p = (p_{source}, p_{target})$ where $p \; \varepsilon \; prev_{level}$, determine all edges $ed = e_{in} \cup e_{out}$ where e_{in} are all the inward edges to p_x and e_{out} are all the outward edges from p_x such that p_x is the node p_{target} if $p \; \varepsilon \; edges_{outward}$ and is the node p_{source} if $p \; \varepsilon \; edges_{inward}$. Afterwards, define a base variable $basis_r = r_p$ where r_p is the corresponding regulation value of the edge p.

Step 5: Define blank sets *subnet* and *regulation*. Afterwards, for each member of $e_x \; \varepsilon \; ed$, determine their corresponding regulation $r_{ex} \; \varepsilon \; r$. If $r_{ex} = basis_r$, and r_{ex} is a complex, or the mechanism behind e_x is binding, then append e_x to *subnet* and append r_{ex} to *regulation*, whenever e_x is a unique member of *subnet*.

Step 6: Save the members of *subnet* and *regulation* to an unordered key-value-pair list $updt = subnet(e_x) : regulation(r_{ex})$. Append a dummy copy of *updt* to a blank key-value-pair list nodules such that $nodules = \{updt\}$.

Step 7: If it is the first iteration, define a set $A = edges(v)$ to hold all the edges connecting to the initial node. Afterwards, define another set $B = subnet(e_x)$ to hold all the edges determined from Step 5 . Define a set $A + B = A \cup B$ and a set $AB = A \cap B$. If not the first iteration, define $A = A + B$ from the previous iteration and $B = subnet(e_x)$.

Step 8: For each member $ab \varepsilon AB$, remove the key ab and its corresponding value r_{ab} from the key-value-pair list nodules. Afterwards, append the value of nodules to *trimNET*. If not the first iteration, redefine $A + B$ to $A + B \cup B$. Afterwards, assign nodules as the new value for $prev_{level}$. If the elements of nodules are null, output the trimmed network *trimNET*. Else append nodules to *trimNET*. Return to step 3.

3.6 Applying the MLST algorithm

From this subnetwork, a modified version of the MCDS algorithm from Dizon & Malolos (2018) was applied. The algorithm was modified by pre-determining the starting node v to a protein we wish to simulate activation or overexpression. Further modification was made by introducing directionality into the MCDS calculation by treating inward and outward edges separately using the directionality indicated by the SIGNOR database.. The resulting MCDS and the overall network was visualized through Cytoscape 3.4.0 (Shannon et al. 2003).

3.7 *Biological analysis*

Gene expression data for breast cancer was taken from CancerNet (Meng et al. 2015). Expression data specific to HER2-positive breast cancer was taken from two sources. The first source is a paper by Ferrari et al. (2016) containing a list of frequently amplified genes in HER2+ breast cancer with information on how frequently this is observed in a set of 64 samples. The second is a paper by Kalari et al. (2013) containing a list of 685 differentially expressed genes in HER2+ tumors with information on the median read count per gene.

The ReactomeFIViz Cytoscape plugin was used to perform pathway enrichment analysis and generate multiple pathway visualizations (Wu et al. 2014). Statistical overrepresentation tests were performed using the online Gene Ontology (GO) tool PANTHER (Thomas et al. 2006). Both Reactome (Fabregat et al. 2015) and PANTHER pathways were used as reference databases for GO analysis.

4 RESULTS AND DISCUSSION

4.1 *Biological evaluation of the directed MCDS*

The initial untrimmed network contained 3892 proteins. Using HER2 as the starting node (denoted as ERBB2 in all Cytoscape visualizations), the trimmed network (trimNET) we generated consisted of 1446 nodes. This network is a simulation of HER2+ breast cancer in the human PPI network. After running the algorithm, we found that the directed MCDS consisted of 333 proteins, or only 8.6% of the untrimmed network. This indicates that the directed MCDS is a small list of highly focused proteins, which makes it ideal for targeted studies.

Cytoscape was an important tool for biological evaluation as it allowed for clear visualization of the directed MCDS network with up/downregulation and gene expression data as explained in Figure 2. Furthermore, using the ReactomeFIViz plugin, pathway enrichment analysis could be performed directly from Cytoscape.

As discussed previously, overexpressed HER2 is known to dimerize with HER3 to trigger oncogenic signaling through either the MAPK pathway or the PI3K-Akt pathway (Moasser, 2007). The directed MCDS identified 19 out of 37 proteins in the HER2/HER3 signaling pathway as determined by the ReactomeFIViz Cytoscape plugin (Figure 3). Visualization of the pathway within the MCDS showed a fully connected subnetwork containing many highly expressed proteins in breast cancer (HER2, AKT1, JUN, etc) as indicated by enlarged node size.

The PRKACA protein was of interest, since it is the only protein in this subnetwork shown to upregulate HER2 (aside from HER2 self-regulation). It is also the only protein in the subnetwork which is part the list of 685 differentially or uniquely expressed genes in HER2+ breast cancer (Kalari et al. 2013) (indicated by its pink node color). PRKACA is known to mediate resistance to therapies that target HER2 as well as sustain anti-apoptotic behavior in tumor cells (Moody et al. 2015). This points to the potential of PRKACA as a possible anti-cancer drug target.

Next, statistical overrepresentation tests were performed on the proteins in the directed MCDS using the PANTHER GO online tool (Ashburner et. al 2000; Mi H et. al 2019; The Gene Ontology Consortium 2019). Overrepresentation tests are important in determining pathways that are more concentrated in a given set of proteins than expected based on their usual incidence in the human genome. Using Reactome pathways as the reference database, the most overrepresented pathways were found to be oncogenic signaling events (Figure 4). These involve phosphorylation, mutation, activation or inhibition events that are specifically linked to cancer.

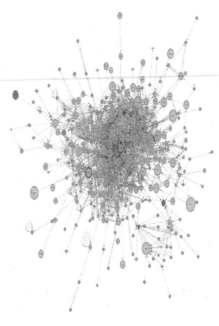

Figure 2. Cytoscape visualization of the directed MCDS using HER2 as starting node. Up and down-regulation are indicated by green and red edges, respectively. Node size reflects the gene expression level in breast cancer. Node color intensity reflects the median expression of the gene encoding for the given protein in HER2+ breast cancer samples. Node border color intensity reflects the percentage of HER+ breast cancer samples in which the corresponding gene is amplified. The directed MCDS contains 333 nodes.

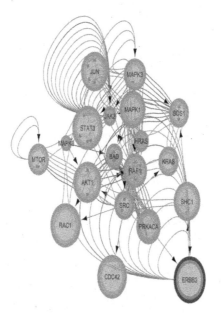

Figure 3. Cytoscape visualization of the HER2/HER3 signaling pathway in the directed MCDS as identified by the ReactomeFIViz plugin. The directed MCDS identified 19 out of 37 proteins involved in the pathway, all of which were connected to each other. Note that HER2 is denoted as ERBB2 in all Cytoscape visualizations.

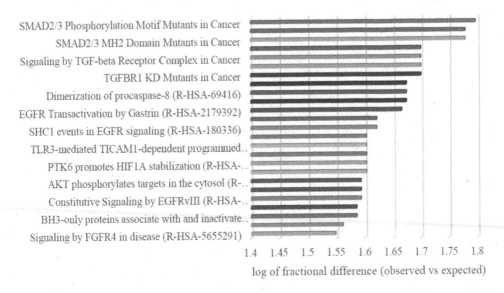

Figure 4. Statistical overrepresentation in the directed MCDS using Reactome pathways from the PANTHER online annotation tool. Fishers Exact test was used with a 0.05 p-value cutoff.

In summary, the findings from pathway enrichment and statistical overrepresentation illustrate the capability of the directed MCDS to well reflect the targeted biological condition. The biological specificity of the directed MCDS will be valuable for studies on key proteins in specific types of cancers or diseases.

4.2 Comparing the directed and undirected MCDS

To assess the uniqueness of the directed MCDS, we also generated the undirected MCDS for the untrimmed network using the algorithm we had prior to our improvements. We inputted HER2 as the starting node for consistency. The generated MCDS consisted of 811 proteins, which is significantly larger and more complex than that of the directed MCDS (see Figure 5). This was expected, since the old algorithm does not contain a network trimming step and instead generates a more general MCDS for the entire network.

Despite the larger size of the undirected network, the directed MCDS identified 68 unique proteins, meaning 20% of the directed MCDS was unique. We ran the list of unique proteins through PANTHER GO analysis to check which pathways they were most involved in. Using PANTHER pathways as a reference, we found that many nodes in the unique set were involved in cancer related pathways such as apoptosis, Wnt signaling, EGFR signaling, and angiogenesis (Figure 6). Interestingly, we also found that the unique proteins were involved in various immune responses such as T cell activation, inflammation, and interleukin signaling. This supports biological studies that have found HER2 overexpression to alter immune pathways (Nicolini et al. 2018) as well as trigger inflammation through interleukin signaling (Liu et al. 2018).

The involvement of unique proteins in immunity became even more evident when we used Reactome pathways as a reference database for PANTHER GO analysis. Cytokine signaling and interleukin signaling were the top pathways for the set of unique nodes (Figure 7).

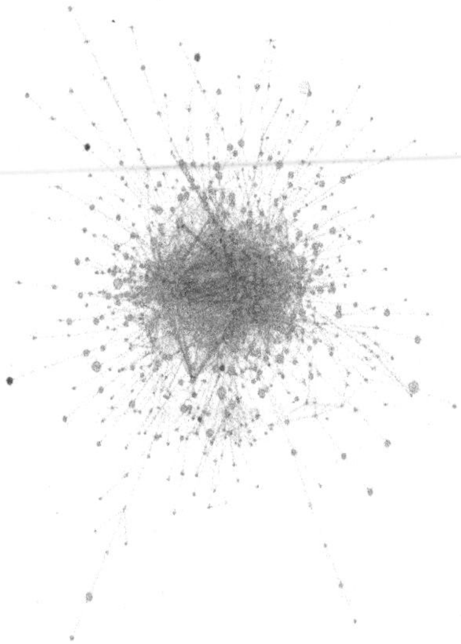

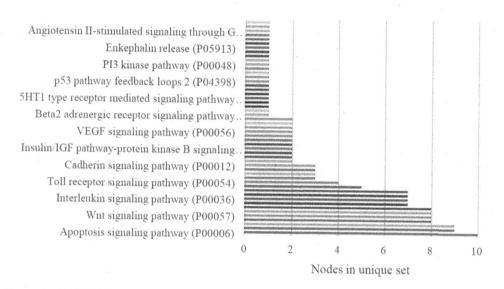

Figure 5. Cytoscape visualization of the undirected MCDS using HER2 as the starting node. Size: 811 proteins.

Figure 6. PANTHER pathway analysis of 68 nodes unique to the directed MCDS using the PANTHER online annotation tool.

The ability of the directed MCDS to identify unique proteins that participate in known side effects of HER2+ breast cancer strongly suggests that it is a powerful tool for targeted biological studies. While the undirected MCDS serves its own function as a general backbone of the entire system, the directed MCDS is likely the better choice to generate a more condition-specific set of key proteins when zooming into a type of cancer or disease. Thus, testing the directed MCDS on different diseases is highly recommended.

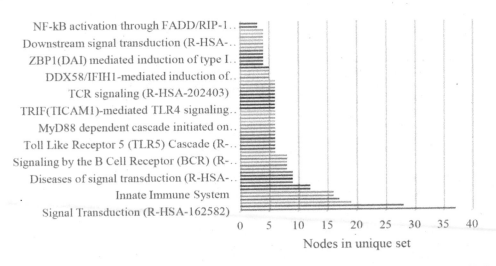

Figure 7. Reactome pathway analysis of 68 nodes unique to the directed MCDS using the PANTHER online annotation tool.

5 CONCLUSION

Protein-protein interaction networks are inherently dynamic, yet their computational analysis has remained static for a long time. This study aimed to develop a dynamic minimum connected dominating set algorithm by modeling it on biological principles of directionality and regulation. Using HER2-positive breast cancer simulation as a test case, the directed MCDS identified 333 out of 3892 proteins in the SIGNOR database. Statistical overrepresentation tests showed that these proteins were highly involved in cancer-related pathways. Furthermore, the directed MCDS identified 68 nodes not present in the larger undirected MCDS. These proteins are involved in cancer and immune system inflammation pathways, which suggests that incorporating directionality results in higher specificity to the given biological condition.

Extensions for this study include exploring other biological conditions by running the directed MCDS using different starting nodes. Furthermore, other databases that indicate directionality may be explored. Finally, the algorithm can be improved by accounting for other parameters in protein-protein interactions such as activation, inhibition, and phosphorylation.

ACKNOWLEDGEMENTS

We thank Mr. Joshua Dizon and Dr. Alexander Young for guiding us with our study. This study was funded by DOST-PCIEERD under the IMBUE project of the Philippine Genome Center.

REFERENCES

Amala, A., & Emerson, I. A. 2019. Identification of target genes in cancer diseases using proteinpro–tein interaction networks. *Network Modeling Analysis in Health Informatics and Bioinformatics* 8(1). doi:10.1007/s13721-018-0181–1.
Ashburner et al. 2000. Gene ontology: tool for the unification of biology. *Nat Genet.* .25(1):25–9.
Barabsi, A., Gulbahce, N., & Loscalzo, J. 2010. Network medicine: A network-based approach to human disease. *Nature Reviews Genetics* 12(1): 56–68. doi:10.1038/nrg2918.
Bhowmick, S.S., & Seah, B.S. 2016. Clustering and Summarizing Protein-Protein Interaction Networks: A Survey. *IEEE Transactions on Knowledge and Data Engineering* 28(3),638–658. doi:10.1109/tkde.2015.2492559.

Das, J. & Yu, H. 2012. HINT: High-quality protein interactomes and their applications in understanding human disease. *BMC Systems Biology* 6(1): 92.

Dizon, J. & Malolos, J. 2018. Minimum Connected Dominating Sets on Protein-Protein Interaction Net-works. CS 297, *Discrete Mathematics*, December 208.

Fabregat, A., Sidiropoulos, K., Garapati, P., Gillespie, M., Hausmann, K., Haw, R., ... & Matthews, L. 2015. The reactome pathway knowledgebase. *Nucleic acids research* 44(D1): D481–D487.

Ferrari, A., Vincent-Salomon, A., Pivot, X., Sertier, A. S., Thomas, E., Tonon, L., Thomas, G. 2016. A whole-genome sequence and transcriptome perspective on HER2-positive breast cancers. *Nature communications* 7. doi:10.1038/ncomms12222.

Fionda, V. 2019. Networks in Biology. *Encyclopedia of Bioinformatics and Computational Biology*: 915–921. doi:10.1016/b978-0-12-809633-8.20420-2.

Goncearenco, A., Li, M., Simonetti, F. L., Shoemaker, B.A., & Panchenko, A.R. 2017. Exploring Protein-Protein Interactions as Drug Targets for Anti-cancer Therapy with In Silico Workflows. *Methods in Molecular Biology Proteomics for Drug Discovery*: 221–236. doi:10.1007/978–1–4939–7201–2_15.

Ivanov, A.A., Khuri, F.R., & Fu, H. 2013. Targeting protein protein interactions as an anticancer strategy. *Trends in Pharmacological Sciences* 34(7): 393–400. doi:10.1016/j.tips.2013.04.007.

Kalari, K. R., Necela, B. M., Tang, X., Thompson, K. J., Lau, M., Eckel-Passow, J. E., Perez, E. A. 2013. An integrated model of the transcriptome of HER2-positive breast cancer. *PloS one* 8(11): e79298. doi:10.1371/journal.pone.0079298.

Kashyap, S., Kumar, S., Agarwal, V., Misra, D.P., Phadke, S.R., & Kapoor, A. 2018. Protein protein in-teraction network analysis of differentially expressed genes to understand involved biological pro-cesses in coronary artery disease and its different severity. *Gene Reports* 12: 50–60. doi:10.1016/j.genrep.2018.05.008.

Liu, S., Lee, J.S., Jie, C., Park, M.H., Iwakura, Y., Patel, Y., Chen, H. 2018. HER2 Overexpression Triggers an IL1î± Proinflammatory Circuit to Drive Tumorigenesis and Promote Chemotherapy Resistance. *Cancer research* 78(8): 20402051. doi:10.1158/0008-5472.CAN-17-2761.

Lo Surdo, P., Calderone, A., Cesareni, G., & Perfetto, L. SIGNOR: A Database of Causal Relationships Between Biological Entities-A Short Guide to Searching and Browsing. *Current Protocols in Bioinformatics*, 2017, doi:10.1002/cpbi.28.

MÄ•nard, S., Tagliabue, E., Campiglio, M., & Pupa, S.M. 2000. Role of HER2 gene overexpression in breast carcinoma. *Journal of Cellular Physiology* 182(2): 150–162. doi:10.1002/(sici)1097-4652(200002)182:23.0.co;2–e.

Meng, X., Wang, J., Yuan, C., Li, X., Zhou, Y., HofestÃ¤dt, R., & Chen, M. 2015. CancerNet: a Database for Decoding Multilevel Molecular Interactions across Diverse Cancer Types. *Oncogenesis* 4(12). doi:10.1038/oncsis.2015.40.

Moasser, M.M. 2007. The Oncogene HER2: Its Signaling and Transforming Functions and Its Role in Human Cancer Pathogenesis. *Oncogene* 26(45): 64696487. doi:10.1038/sj.onc.1210477.

Moody, S. E., Schinzel, A. C., Singh, S., Izzo, F., Strickland, M. R., Luo, L., Hahn, W. C. 2015. PRKACA mediates resistance to HER2-targeted therapy in breast cancer cells and restores anti-apoptotic signaling. *Oncogene* 34(16): 20612071. doi:10.1038/onc.2014.153.

Nazarieh, M., Wiese, A., Will, T., Hamed, M., & Helms, V. 2016. Identification of key player genes in gene regulatory networks. *BMC Systems Biology* 10(1). doi:10.1186/s12918-016-0329–5.

Nicolini, A., Ferrari, P., Diodati, L., & Carpi, A. 2018. Alterations of Signaling Pathways Related to the Immune System in Breast Cancer: New Perspectives in Patient Management. *International Journal of Molecular Sciences* 19(9): 2733., doi:10.3390/ijms19092733.

Parthiban, N., Rajasingh, I., & Rajan, R. S. 2015. Minimum Connected Dominating Set for Certain Circulant Networks. *Procedia Computer Science* 57: 587–591. doi:10.1016/j.procs.2015.07.401.

Purohit, G.N., & Sharma, U. 2010. Constructing Minimum Connected Dominating Set: Algorithmic Ap-proach. *International Journal on Applications of Graph Theory In Wireless Ad Hoc Networks And Sensor Networks* 2(3): 59–66. doi:10.5121/jgraphoc.2010.2305.

Shannon, P., Markiel, A., Ozier, O., Baliga, N. S., Wang, J. T., Ramage, D., Ideker, T. 2003. Cytoscape: a software environment for integrated models of biomolecular interaction networks. *Genome research* 13(11): 24982504. doi:10.1101/gr.1239303.

Simonetti, L., Cunha, A.S., & Lucena, A. 2011. The Minimum Connected Dominating Set Problem: Formulation, Valid Inequalities and a Branch-and-Cut Algorithm. *Lecture Notes in Computer Science Network Optimization*: 162–169. doi:10.1007/978-3-642-21527-8_21.

Skwarczynska, M., & Ottmann, C. 2015. Protein protein interactions as drug targets. *Future Medicinal Chemistry* 7(16): 2195–2219. doi:10.4155/fmc.15.138.

The Gene Ontology Consortium. 2019. The Gene Ontology Resource: 20 years and still GOing strong. *Nucleic Acids Research*, Volume 47, Issue D1, Pages D330D338, https://doi.org/10.1093/nar/gky1055

Thomas, P.D., Kejariwal, A., Guo, N. Mi, H., Campbell, M.J., Muruganujan, A., & Lazareva-Ulitsky, B. 2006. Applications for protein sequence-function evolution data: mRNA/protein expression analysis and coding SNP scoring tools. *Nucl. Acids Res* 34: W645-W650

Tian, J., & Ding, H. 2013. Solving Minimum Connected Dominating Set on Proper Interval Graph. *2013 Sixth International Symposium on Computational Intelligence and Design*. doi:10.1109/iscid.2013.25.

Wu, G., Dawson, E., Duong, A., Haw, R., & Stein, L. 2014. ReactomeFIViz: a Cytoscape App for Path-way and Network-Based Data Analysis. *F1000Research* 3:146., doi:10.12688/f1000research.4431.2.

Zhang, X., Ou-Yang, L., Zhu, Y., Wu, M., & Dai, D. 2015. Determining minimum set of driver nodes in protein-protein interaction networks. *BMC Bioinformatics* 16(1). doi:10.1186/s12859–015–0591–3.

Theory and Practice of Computation – Nishizaki et al (eds)
© 2021 Taylor & Francis Group, London, ISBN 978-0-367-41473-3

Author Index